How Life Is Different

How Life Is Different

Vitaly Shubin

CRC Press is an imprint of the
Taylor & Francis Group, an **informa** business

First edition published 2021
by CRC Press
6000 Broken Sound Parkway NW, Suite 300, Boca Raton, FL 33487-2742

and by CRC Press
2 Park Square, Milton Park, Abingdon, Oxon, OX14 4RN

Library of Congress Cataloging-in-Publication Data
A catalog record has been requested for this book

ISBN: 9780367775551 (hbk)
ISBN: 9780367771812 (pbk)
ISBN: 9781003171898 (ebk)

Typeset in Times
by KnowledgeWorks Global Ltd.

To my unforgettable mother

Contents

Preface xi

Chapter 1 Cell and Unicellular Organisms 1

What Is the Difference between Chemical Laboratories in a Cell and Volcano? 1
Stochastic Process and a Predictable Program 2
Informational Organization of Living 2
Program Parts and Components 3
Pool of Basic "Production" Cell Programs 5
Programs That Produce Makers 5
Provision and Energy Programs 11
Reproducing Sites 14
Transport Programs 15
The Main Objectives of a Cellular Program System (CPS) and Its Important Programs 19
Control Programs 20
Signal Programs 24
The General Cellular Program and Its Most Important Component – Mitosis 27
Cell Cycle Management 29
Other CPS Programs 33
External Programs of Unicellular Organisms 33
A Full Program System of Unicellular Organisms: Management Principles 35
Evolution of Unicellular Organisms 39
Program-Information Cell Structure 40

Chapter 2 Multicellular Organisms 51

Socialization Is the General Principle of Structuring a Species Community 51
Transition from Unicellular to Multicellular 51
Aggregate Organisms 52
Basic Social Principles of Cellular Integration into a Single Structure 53
Program Scheme of a Multicellular Organism (MO) 53
Social Multicellular Organism Structure 53
MO Tasks and Program Systems That Solve Them 54
Internal and External Programs 55
Main and Auxiliary Task Programs 55
Structure of MO Organs and Systems 55

Examples of Internal Programs....56
Nutrition Programs....56
The Main Features of the Central Nutrition Program (CNP)....59
Breathing Programs....61
Immune Programs....65
Tasks and Structure....65
Implementation Examples....66
Some Characteristics....70
Transport Programs....71
Blood Supply System – Basic Transport System of MOs....71
Heart....72
Vessel Structure and Types....72
Circulation....75
The Main Program Features of the Blood Supply System....75
Management Programs....78
Humoral System....79
Nervous System (NS)....80
Neurons Form NS Signal Chains....80
Signal Transfer Program....80
Returning the Neuron to Its Original State after a Signal Transfer Program....81
Program of Signal Transmission between Neurons....82
Program to Restore Neuron Activity....82
Features of Neural Chain Links....83
Input Sensor Links....83
Signal Processing Programs Run by Interneurons and Effector Neurons....84
Effector Neurons....86
Network Structures Formed by Neurons....87
Memory of Neural Networks....93
External Programs....95
External Nutrition Programs....95
Program Space (PS)....97
Structure of External Programs....98
Sensor, Analytical and Effector Blocks of External Programs....99
Work Scheme and Interaction of Blocks....99
Possible Principles of Formation and Inheritance of External Programs....103
Program and Information Structures of Multicellular Organisms....104

Chapter 3 Man....111

Leap Forward in Evolution – Emergence of Human Volitional Programs....111
Social Structure of Human Communities – Personal and Collective Programs....113

Autonomous System of Volitional Programs (ASVPs) 113
Human Social Paradigm – The Relationship between Personal and Collective Programs 114
Types of Collective Programs 114
Allocation of Collective Program Results: ASVP Management and Synchronization 116
About Information in Volitional Programs 122
Memory 123
Collection, Formation and Processing of Information 126
Information Transfer – Communication 126
Structure of Volitional Program Makers 127
History of Mankind as Evolution of Volitional Programs 127
Scientific Stage in Evolution of Volitional Programs 132
Applied and Fundamental Science 133
Scientific Methodology 133
Use of Space and Time Concepts 135
Aspects of Time Concepts in Program-Information Approaches 136
Stochasticity at Inanimate Nature 138
Combating Stochastic Degradation in Program Systems 139
About Learning 140
About the World Picture 141
Structure and Principles of Volitional Programming – Basic Elements and Operations 144
Language and Volitional Programming 144
Images and Symbols 145
Hierarchy of Personal Information Programs 147
Possible Schematic Scenarios of Stimulation and Manipulative Programs 148
Inheritance and Development of Unicellular and Multicellular Information Systems in Volitional Programs 149

Chapter 4 Program-Information Unity of Living 153

Common Program and Information Structures 153
The Essence of Cellular Evolution 154
Cell and Unicellular Organisms 155
Makers 155
Program Structure 156
Management of Programs 157
Information Network 157
Tasks and Structure of Information Programs 158
Complete System of Unicellular Programs 159
Cellular Evolution 161
Multicellular Organisms 161
Structure of Multicellular Organisms 161
Tasks of MO Programs 162

Management of MO Programs 163
Nervous System 164
Features of the Information System 165
Memorizing and Learning 166
Improvement of External Programs at the Highest MO Representatives 167
Man 167
Brain Evolution and Hand Motility – Basis of Homo Programming Appearance 167
Social Nature of Human Communities – ASVP Occurrence 168
Personal and Collective Volitional Programs 168
Social Allocation of ASVP Resources 169
Management in ASVP 170
Coordination and Synchronization of Programs 171
Types of Management Regulatory Programs 172
History of Evolution of Volitional Programs 172
The Role of Science 173
The Scientific Method from the Standpoint of the Program Approach 174
About the World Picture 174
Mechanisms of Volitional Programming 175
Features of Volitional Information Systems 176
Evolution and Program Approaches 177

Provisions of the Program Approach 179

Reading List 187

Index 189

Preface

Although it is traditionally believed that getting answers to the questions: "What is life?" and "How did life come about?" is a topical task of science, there are not many essential works on this subject. Apparently, this can be explained by the fact that the representatives of the natural sciences are busy solving their "urgent" fundamental and applied problems, and philosophers do not fall to the level of concrete consideration of the problem.

Among the works devoted to this topic, we can mention the works of Schrödinger, Brillouin, Quastler, Blumenfeld, Chernavskii, Ivanitskii and others.

All these works combine attempts to explain the living through the fundamental laws of the non-living, in particular, using a statistical approach that has fruitfully proved itself when considering the non-living multi-element molecular, atomic, electronic sets. This approach is based on the initial consideration of the ideal gas.

The statistical set of identical equivalent interacting objects – molecules, is considered. Further, the assumption of "molecular chaos" is implied as stochastics, leading to uniform distributions of molecules in the space under consideration and their velocities and energies in all directions of this space. When describing the behavior of other ensembles, not only molecular objects, the main elements of the approach – the same equivalent objects and stochastic reproduction of their observed parameters remain.

The approach allows us to consider the dynamics of the development of stochastic processes in the form of transitions from one equilibrium state to another.

It should be noted that the statistical approach has been successfully applied to individual situations with living objects when sets of their local signs and characteristics are considered, for example, in studies of some features of the behavior of biological systems in microbiology and ecology.

However, it must be admitted that the number of such successful use of statistical methods in the study of living objects is small and has no direct relationship to solving basic issues of organizing living.

Most scientists dealing with the general problems of the structure and origin of life have a dissatisfaction with their purely physical explanation and, accordingly, the use of a statistical approach for this. However, they all agree on the unconditioned microscopic physical nature of elements and their interaction in living systems. But the real complex, organized, expedient structure and behavior of organisms requires, in their common opinion, the introduction of some kind of globally new quality in living systems, compared with non-living ones.

Their statements on this subject can be summarized as follows.

Schrödinger noted that the unfolding of events in the life organism cycle of the body reveals amazing regularity and ordering, which is unparalleled among everything that we encounter in inanimate matter. And then he adds that in the study of living objects, we encounter phenomena whose regular deployment is determined by a "mechanism" that is completely different from the "probability mechanism" of physics.

Blumenfeld writes that the usual thermodynamic approach is in most cases unproductive when considering living systems and their components, and in particular one should not expect much from measurements of the special ordering of biostructures in entropy units.

Finally, Quastler believes that in the normal course of events, orderliness tends to decrease, and therefore it is not easy to understand how living things could arise from inanimate predecessors.

The author of this text builds his concept of organizing life intentionally and, as he hopes, reasonably, without using a statistical approach.

The book *How Life Is Different* is based on two main issues:

- "What is the difference between the interaction of bodies and objects in an animate and inanimate world?"
- And: "What is the mechanism that implements this difference?"

An analysis of these issues at the beginning of chapter—"Cell and Unicellular Organisms," on the material of events and processes occurring in the cell demonstrates a cardinal difference between regulated, ordered interactions involving proteins and ribonucleic acids from stochastic interactions in the inanimate world.

First, participants in biological interactions are not equivalent. The active participant (hereinafter referred to as the maker, a protein or its complex usually) performs a priori certain actions with respect to the passive object. Such actions can be: mechanical, for example, moving an object, dividing it into parts, connecting the parts of the object into a single whole; chemical, changing the chemical state of the object, for example, phosphorylation, methylation and/or mixed chemical-mechanical.

Further, there is a priori uniqueness of the result of such an action, i.e., changes in the state of an object. This is due to the specific reproducible structure of the maker that recognizes and selects only its object for exposure and performs a certain strictly regulated action in relation to it.

The result of a "live" interaction is a change in the state of the object while maintaining the structure of the maker for the following interactions in the presence of exactly the same object and energy supply, for example, in the form of ATP. The "meeting" of the interaction components: the maker, the object and the energy, takes place at a specific site inside specific cellular compartments or at sections of their membranes.

The combination of actions of makers on one or several objects makes up a regulated program with an a priori predicted result of task solving, for example, creating a protein, producing an ATP energy molecule, building up cell compartment membranes, transporting makers and objects to the places (sites) of their interaction, and finally, a general cellular program cell division.

The process consisting of stochastic inanimate interactions does not end with an a priori predicted result and generally leads to an increase in the entropy of the aggregate of interacting objects, which we, in particular, observe after the death of the organism, with the aggregate of the same proteins and RNA which participated in live programs before that.

"Living water", which turns an inanimate stochastically interacting set of atoms and molecules into living programs, is information – a system of instructions and signals, in accordance with which the programs are reproduced and consistent with each other. The information, in turn, is the result of the implementation of special information programs that are part of the complete program system of the cell and any organism, along with the above-mentioned task effector programs.

The most important information programs determine, in particular, reproduction of makers specialized for a given object, by DNA and RNA instructions; identification of objects and their delivery addresses using the maker's information domain. In addition, signal-information management of the organism program system is carried out by optimizing the frequency and intensity of the execution of individual programs and their groups.

Important makers of information programs are the receptors that record external and internal situational signals, and the "deciders" who decide on the basis of these signals how to adequately adjust the organism state to this situation.

The complete program system of the cell and any organism (called the program type or briefly protype) ensures its reproduction and adaptation to external conditions, including evolutionary over long periods of time.

We can say that a protype based on a genotype creates an organism phenotype.

The book contains four chapters "Cell and Unicellular Organisms", "Multicellular Organisms", "Man" and "Program-Information Unity of Living". The "Provisions of the Program Approach" provides the final conclusions of the book.

Thus, the author, in addition to formulating a program approach using the example of a cell, tried to apply it to the description of subsequent evolutionary stages – multicellular and human, and then to generalize the consideration results of the first three chapters in the final fourth chapter.

In the chapter "Cell and Unicellular Organisms" on the material describing the cell's structure and the events occurring in it, the basic concepts of an ordered, regulated program carried out by active interaction participants – makers with respect to passive objects using energy particles (ATP) on the certain cell compartment sites.

The chapter then discusses specific task programs for creating makers, storing energy, reproducing compartments, etc., as well as the general cell cycle program. A complete system of internal and external cellular programs that form the so-called "program type" or "protype" of the cell is presented.

The chapter "Multicellular Organisms" discusses the system of social programs carried out by makers of different hierarchical levels: elementary intracellular, cellular, tissue and organ makers. These programs perform the basic tasks of nutrition, respiration and immune defense against microorganisms attacking the body. The role of the nervous system in the implementation of control programs for internal and external behavioral program groups is shown.

The chapter "Man" is based on the introduced concept of volitional programs. This term refers to the complex of external programs of a person that he is able to independently compose, modify and execute due to the evolutionary appearance in his brain of the Wernicke and Broca fields, responsible for the perception and processing of the language.

The linguistic secondary image system made it possible, first, to formulate and virtually reproduce program elements, their script, to track and evaluate the progress of their implementation and, second, to carry out communication between people for coordinated and often synchronized collective execution.

The book discusses the topics of personal and collective volitional programs, methods of their management and coordination, historical and evolutionary development, as well as elements of the individual mechanism of volitional programming.

Using the proposed program approach, according to the author, is possible in different directions.

The most important area of research in modern biology is the establishment of a correspondence between an organism phenotype and its genotype.

From the standpoint of a program approach, there is the intermediate linking the genotype with the phenotype. These are the programs that implement phenotype signs in accordance with gene instructions.

A gene or their combination does not create phenotypic characters, but merely serves as an informational instruction for transcription – the first stage in the creation of makers: RNA, proteins and their complexes to carry out the necessary programs.

In addition to the makers, for the successful implementation of these programs, it is necessary to provide them with the objects, energy in required concentrations on the corresponding sites.

Thus, in order to effectively clarify the relationship between the genotype and phenotype and, moreover, to implement the required phenotypic characters, the detailed consideration of the programs that implement the signs is necessary.

The compilation of a cell program passport reflecting the system of its programs could play a significant role in identifying, for example, the difference between a cancer cell and a healthy one. The effect on a cancer cell is possible both at the level of correction of its individual programs or their groups, and by reconfiguring or modifying its control program unit.

The use of the program approach when considering physiology and developing therapeutic methods in medicine also seems potentially fruitful.

Anthropology, psychology and sociology can be referred to disciplines more remote from biology for which the use of the program approach is possible.

For all three of the mentioned scientific branches, it may become productive to consider a complex of human volitional programs.

In the case of anthropology, this is a program classification of the activities of *Homo sapiens*, both modern and historical.

Sociology can use the consideration of personal and collective volitional programs, their management and synchronization.

Finally, for psychology, questions of the individual mechanism of volitional programming discussed in the book may be interesting.

As you can see, issues of various specific areas of knowledge are addressed in the book. Readers with different directions and levels of background may be interested in it.

The author is far from thinking that the book presented will become a bestseller, but it would be very desirable for people of differentiate creative, labor and scientific specializations to read it.

Focusing on such a wide and diverse readership, the author tried to solve contradictory tasks: on the one hand, it was to look convincing to specific specialists in the areas in question, and on the other, to be intelligible for the rest.

Of course, it was not possible to completely solve these problems; therefore, the author encourages specialists to be patient and others, on the contrary, to be more active and bolder when encountering new terms and materials when reading the book.

This is especially true for the various scientific names of proteins, enzymes and makers introduced by the author. It is impossible to do without them when describing specific programs.

The Internet completely removes possible problems. Half a minute after entering the seemingly intricate designation of the object in the search engine, you have the necessary and sufficient idea about it to continue meaningfully reading.

When reading a book, representatives of different fields of knowledge can be recommended an "economical" partial reading of the book, which in this case is advisable to start with the general summing up of Chapter 4, "Program-Information Unity of Living"; and then to move on to specific materials of interest to them. For example, anthropologists and psychologists might read Chapter 3 "Man", immunologists, physiologists and ethologists – Chapter 2 "Multicellular Organisms". As for molecular biologists and microbiologists, their "minimum" is enclosed in Chapter 1.

Finally, for the "fans" of the problem as a whole, but at the same time for the "healthy" pessimists regarding the new approaches, we can recommend reading "backward". First, at the end of the book, "Provisions of the Program Approach" containing the main conclusions of the book, and the we warm up and even become annoyed by their non-traditional nature. Then we proceed (with moderate irritation) to the short concluding chapter, "Program-Information Unity of Living", the reading of which should, with its simplified system of arguments, reduce irritation and increase interest in the subject. And then, at a favorable outcome for the author, we begin to read the contents of the three main chapters with a mixed feeling of distrust and interest.

1 Cell and Unicellular Organisms

WHAT IS THE DIFFERENCE BETWEEN CHEMICAL LABORATORIES IN A CELL AND VOLCANO?

Judging by the title of the book, the author is going to express his thoughts on the structure of life, specifically, on the living part of the observed world.

Because life is a collection of organisms and their vital functions, and the body is just set of cells, the cell is the basis of all life systems.

Therefore, in order to answer the question of the chapter, we need to examine what happens in the cell.

In the cell, there are plants and laboratories that produce products and carry out operations for the growth and functioning of the cell throughout the entire cell cycle. Moreover, the words factories and laboratories should be used without quotation marks, because they are much better in organization, efficiency and product quality than the most modern human plants and laboratories, furthermore we can take into account the minimization of working space and energy costs as well as strict quality control indicators in cell plants and factories.

Indeed, we humans have a lot to learn, for example, in assembly production, taking a protein assembly as a model, or in the process of organizing transport by observing various cellular solutions to the problem of delivering large and small molecules and ions to the right places in the cell compartments, The processes of management of both individual operations and the functioning of cellular systems and the whole cell also deserve close attention.

It should be noted that this particular work of cellular plants is being thoroughly studied and quite successful, but the examination of the system's structure is somewhat lagging behind.

Still, the main processes happening in the cell are considered biochemical reactions of decomposition, catabolism and synthesis, anabolism, carried out with the help of enzymes which are mainly proteins and complexes based on them. This is so, but they do not explain rather striking differences between a living cell laboratory and non-living one, for example, in a volcano.

Why not to compare them? In both cases, we observe chemical reactions and mechanical movements of interacting particles and bodies as well as changes in their forms and characteristic features. Also, the presence of analogues of enzymes – catalysts in a volcanic laboratory is not excluded.

However, the difference among the laboratories is enormous.

Systematic, regulated, that is precise in time and space activity in the cell leads to exact reproducible results and stochastic processes occurring in random time and space, lead to result of the emergence of unexpected gifts, such as porous tuffs with different colors, or glassy black obsidian in a volcanic laboratory.

Well, we have touched, as we see it, the fundamental differences between the stochastic process in the inanimate world and the regulated live process, hereinafter referred to as the program.

Let us try to give definitions of both, as well as related definitions.

STOCHASTIC PROCESS AND A PREDICTABLE PROGRAM

The process – at least one equivalent, interchangeable interaction of two arbitrary bodies or objects, occurring at an indefinite time in an indefinite place, accompanied by the conservation of energy while taking into account its dissipation and leading to the result.

The result of the process and its consequence is determined by the characteristics of two interacting bodies and their interaction and may consist in changing the spatial parameters of the bodies: acceleration, speed, position (coordinates) and/or changes in their characteristics: size, shape, chemical state, etc., including combining two interacting objects into one or dividing them into several objects.

The result of the process as well as the process itself is not predetermined in advance, i.e., it is random, stochastic both in essence and in place and time of its realization.

The program is at least one predetermined, regulated action made by a specific performer (maker) with respect to a predetermined object at a predetermined time in a predetermined place in space – the program execution site, mainly accompanied by the absorption of energy stored by special energy programs (EPs).

The result of the program may consist in changing the spatial parameters of the object: acceleration, speed, position (coordinates) and/or changes in its characteristics: size, shape, chemical state, etc., including dividing the object into several objects or combining the object and the performer, and, in accordance with the definition of the program, it is predetermined in its essence, time and place of implementation.

An important clarification! Here we are considering the initial natural process, which has not been subjected to human study, and all the more so to the impact, which is based precisely on the program approach. In other words, the result of observation and study of the initial stochastic non-living process bears the subjective traces of the program approach. The essence of the program approach in science will be discussed in more detail in the third chapter of this book.

INFORMATIONAL ORGANIZATION OF LIVING

First, we must note that a large part, and perhaps all actions of performers over an object by nature or as we will speak further on modality, correspond to the interactions observed in inanimate nature. These are primarily mechanical and chemical interactions, leading to movement, change in shape and separation into parts, association, change in chemical activity, etc., in bodies and objects involved in interactions. At the same time, the material molecular, atomic and field basis of the program living world, completely match with that for its inanimate part.

So what kind of life-giving water turns the stochastic inanimate process into a living and ordered program?

Such life-giving water is information – a system of instructions and signals, in accordance with which programs are reproduced and are consistent with each other. The information in turn is the result of the implementation of specific information programs that are part of the program system of the cell and any organism.

The basic part of the information system (IS) is the instructions according to which the performer (maker) recognizes his object and/or performs the necessary program steps and actions.

The most important example of instruction is information of the DNA gene, according to which special programs reproduce blanks for the manufacture of all the makers – ribonucleic acid (RNA) and proteins.

A common example of informational signals are ions or molecules that activate or inactivate a given maker, thereby controlling the frequency or intensity of its execution of a program in accordance with the needs of it for the entire cellular program system (CPS) or its part.

At the same time, in addition to effective maker domain responsible for the implementation of regulated actions on an object, there are information domain (ID) for reading instructions and responding to signals.

We can say that the makers created by the information instruction, performing their actions with regard to the information signals, reproduce all the cell programs and ensure the functioning of the CPS.

Thus, the information organization permeates the entire program structure, its functioning and forms programs.

PROGRAM PARTS AND COMPONENTS

The main components and elements of the program are maker, object, action, and site.

We remind you that we are considering only the cell.

The main active elements of the programs are makers. They are proteins, RNAs and assemblies created on their basis, which are also created as a result of executing programs, the key of which are transcription and translation programs performed by complex makers – polymerase and ribosome, respectively.

Why, instead of introducing the new term "maker," didn't the author use the traditional term "enzyme," which would be, in its own way, logical, since we are talking about the same biochemical formations?

The reason lies in the fact that the author sees in these well-known biochemical objects new, non-traditional functions and capabilities that are fundamental for the developed program concept.

First, this is a significant expansion of the actions performed by the maker in comparison with the enzyme in its classical definition and presentation, in transport, assembly, signal and other programs.

Second, this is the most important property that allows the maker, the performer of its basic program to participate additionally in information programs as an active performer, for example, when reading instructive DNA information, or a passive object when reacting to the signal of the control program.

The maker, generally, performs program actions in relation to a single object, i.e., the structure and functioning of each maker are very specific, which is reflected in the diversity of protein molecules and the complexes formed from them, even in the simplest prokaryotic cells.

The objects of programs can be a variety of cellular components: large and small molecules, atoms, ions and including the makers themselves, for example, as already noted with control programs.

The changes that have occurred with the object are the result of the program, while the maker does not change and after the execution of the next program, he is able to perform similar program once more.

We should mention some more properties and features of the makers and their variants.

At the lower level of the program hierarchy there is a principle: one program – one maker – one object. It corresponds to the situation when a simple maker performing single action on an object, for example, one stage of catabolism, execute a simple program that have an intermediate result.

Other simple programs that are also performed by simple makers use this result. A chain of such simple programs is formed, at the output of which the main, final result is achieved such as, for example, the Krebs catabolic cycle. In this case, we can speak of a complex result program consisting of components of simple programs that are executed sequentially by several simple makers.

In turn, composite programs can be combined into task programs to solve specific cellular problems, for example, breathing.

In the future, we will adhere to such hierarchical organization of programs. Task programs, often executed in different sites and receiving the final result, consist of composite programs that implement intermediate results, usually in the same site, but using a number of makers. Composite programs include program components executed by a single maker.

Both composite and task programs can be formed from the programs of the lower hierarchy by a sequential, parallel or mixed principle.

In the considered variants of programs, simple programs with simple makers performing one action were meant at the lowest hierarchical level.

However, there are well-known makers who perform several actions. For example, the already mentioned ribosome in the translation program performs a series of actions, being essentially a multifunctional complex maker of a composite broadcast program.

Some of the simple one-process programs are performed by different sections of the ribosome, and some, for example, the delivery of amino acids for the formation of a peptide chain, are performed by independent makers – transport RNA.

Thus, we can say that there is a variety of multifunctional, complex makers that perform composite, task and implementing the final result types of the program. In addition to the ribosome, these include RNA polymerase or adenosine triphosphate (ATP) synthase. In some cases, the maker works with two or more objects.

Auto-assembly programs, for example, microtubules or complexes of proteins and RNA, the author considers as programs with a maker and an object. In this case, the role of the maker is performed by the next component that is included in the assembly, and the object is the growing seed of the assembly. Maker identifies the place for assembly and sits on it, fulfilling the task of further formation of the object.

The main actions of the performers can be mechanical, chemical and mixed. Mechanical actions help the movement of the object, chemical actions create common reactions such as oxidation, reduction, phosphorylation and others. Mixed actions lead to the mechanical unification or separation of objects, the mechanism of which is most often concern with the weakening or strengthening of the chemical interaction among individual objects or parts of one object.

Most programs require certain amounts of energy. Typically, energy amounts transmit to makers or objects "energy particles" of ATP.

Localization of the execution of programs is achieved by the presence of various membrane structures of the cell which create its organelles.

Programs can be carried out both inside organelles and directly on the membranes which form them.

Cell membranes, which are based on lipid layers, are assembled by appropriate programs.

Special transport programs collect into according sites all the necessary missing components of the programs running on these sites.

In a number of the following sections of this chapter, instead of the terms "maker", "performer", and sometimes "the working element of the program", we will use traditional definitions such as enzyme, protein, transport protein, synthetase, kinase, etc., implying that any program is always performed by the maker.

POOL OF BASIC "PRODUCTION" CELL PROGRAMS

In order for every program in each cell's life moment to have all the necessary components: makers, objects, energy and sites their reproducing programs included in the pool of "basic" programs should work intensively. To this base pool, we also add transport programs that ensure the delivery of components and their elements to sites and call the programs of this pool – production programs. Let us consider the examples of programs in this pool in more detail.

We will begin our consideration with the programs that make performers or as we will call them makers. This is primarily due to the importance of these programs, and their key position in the cellular system of programs. In addition, using their example, it is convenient to get acquainted with the main features of the structural organization of programs, their interconnections and control elements in order to prepare a subsequent consideration of the totality of all cellular programs as a single system.

PROGRAMS THAT PRODUCE MAKERS

As we have already mentioned, the makers in the cell are RNAs, proteins and complexes of them, which is the result of the corresponding task programs.

At the first stage of all these programs, compound transcription programs are performed, when different types of RNA molecules are created using gene DNA instructions, and their characteristics vary depending on the types of makers created further on their basis. The main types of RNA: matrix messenger RNA (mRNA) which carry information for protein synthesis, transport transfer RNA (tRNA) that are performers implementing a program to deliver the required amino acid to the ribosome, collecting peptide chain, and ribosomal RNA (rRNA) that make up the ribosome.

The transcription programs are performed by three types of RNA polymerase makers, each of which is responsible for the transcription of various types of RNA: RNA polymerase I for rRNA, RNA polymerase II for matrix mRNA and RNA polymerase III for transport tRNA.

All types of transcription program results go through the stages of the so-called post-transcriptional transformations, the results of which are: tRNA maker, a component of the ribosomal assembly, ready to execute its program, rRNA, which is not an independent maker, and, finally, matrix mRNA, which is an RNA instruction for the production of peptide chains of all cellular proteins in translational programs.

Thus, after performing two consecutive programs, which are transcription and post-transcriptional processing, the RNA chains obtained and in only one case the final result is achieved in the form of a ready maker – transport RNA. If it is achieved the task program will consist of the two above-mentioned composite programs.

rRNA and mRNA are intermediate results used to perform subsequent composite programs in task programs with the goal of producing the makers, such as the ribosomes, all protein makers and other complex makers using proteins.

The central component program in task programs to produce protein is translation – a program for assembling a peptide chain according to mRNA instructions.

It is followed by programs that produce a post-translational modification of proteins, of which the most important are the folding programs – imparting the protein chain with the required spatial configuration, performed by the makers-chaperones.

At last, the final stage of manufacturing complex compound makers, the ribosome, polymerases, and others – assembly of RNA and proteins is carried out.

Let us consider as an example two task programs for the manufacture of makers: a short one, the result of which is transport RNA and a long one which creates ribosome for the peptide production.

As we have already discovered, the first things that gets executes in both cases are transcription programs.

In the first case, the main maker of this program is RNA polymerase III, which produces a transcript of the transport RNA. In the second case, at the first stage of a long task program, two transcription programs must be executed. One of these, performed by RNA polymerase II, produces messenger mRNA, which serves as an instruction for a subsequent translation program that collects peptide chains for proteins forming the ribosome. The other is performed by RNA polymerase I and manufactures rRNA, another major component of the assembled maker, the ribosome.

Let us consider the execution sequence of the constituent programs and individual actions, or, as we will often call it, the script of a transcription program using the example of *Escherichia coli*. The script consists of three sequentially executed constituent programs: initiation, elongation and termination.

The compound maker of this program, bacterial RNA polymerase, is formed from several subunits, which at different stages of transcription attach to the polymerase-DNA complex, and then leave it.

The cell nucleus is the transcription site, in which there are elements of the complex maker – its base RNA polymerase and auxiliary protein subunits, instructive information object – DNA, ribonucleotide triphosphates – RNA nucleotides, which constitute the resulting RNA chain and ATP energy molecules used by the maker to perform program actions.

At the first stage, called initiation, recognition of the first instructive information object of the transcription program, the gene section of DNA, which is than transformed into an RNA chain is performed.

To perform the first stage, the protein sigma-subunit (σ) is attached to the RNA polymerase, which allows the resulting maker-complex to find and recognize the promoter nucleotide sequences preceding the DNA gene, according to the instructions of which it is supposed to carry out the synthesis.

When initiating, the enzyme protein complex RNA polymerase – (σ) -subunit, while binding to the DNA with weak bonds, slides along until it binds tightly to the characteristic nucleotide sequence of the desired promoter.

For the active implementation of the subsequent stages of transcription – elongation, in other words process of increasing the RNA chain and then terminating this program, it is necessary that the initiation factor – (σ) 1/m subunit separates from the enzyme complex and other proteins join, for example, nusA protein which acts as an elongation factor.

At the elongation stage, the updated RNA polymerase complex performs a step-by-step process of simultaneous unwinding, melting of the DNA double helix, after which one of the ribonucleotide triphosphate that is capable of complementary pairing with an open DNA chain is selected and then, this nucleotide is connected to a growing RNA strand. After the next advancement of RNA polymerase, the previous RNA link is detached from DNA and the DNA is annealed in this place. As a result, a growing RNA chain and a reconstructed double helix of DNA remain behind a moving RNA polymerase.

The transcription program is executed until the RNA polymerase complex reaches a specific nucleotide sequence in the DNA chain, producing a termination signal, stopping transcription (stop signal). Reaching this point, the polymerase separates from both the DNA and the newly synthesized strand of RNA, called the primary transcript.

The management of the transcription program of the first program in a series of all task programs for the manufacture of makers is crucial for the restructuring of the CPS, in particular, at different stages of the cell cycle.

There are various information programs that regulate the program of transcription of genes, the program is chosen depending on the need for specific makers at varying intensities of their programs.

Eukaryotic cells contain many site-specific DNA-binding proteins, the main function of which is to positively or negatively regulate the transcription of the corresponding genes. Each of these regulator protein makers recognize specific DNA sequences located in front of the gene and then get localized on DNA near this region.

An example of negative regulation is the inhibition of the transcription of the lac-operon of an *E. coli* bacterium consisting of three genes, which create three makers that are necessary for lactose transport and metabolism.

In this case, the repressor protein binds to the so-called operator – a DNA sequence with 21 nucleotides in length, which overlaps with the adjacent promoter RNA polymerase binding site. If the repressor remains linked to the operator, access of the RNA polymerase to its binding site is closed and transcription of nearby DNA regions does not occur.

The lactose repressor in turn is under the control of a small informational molecule of allolactose, which is formed in the cell in the presence of lactose. When the intracellular content of allolactose reaches a sufficiently high level, it, acting as an allosteric regulator and induces conformational changes in the repressor molecule. At the same time, its connection with DNA is weakened to such an extent that it is separated from it, freeing the promoter and allowing the RNA polymerase to transcribe the adjacent regions of DNA. In this case, we can describe this process like the derepression of the gene.

An alternative method of transcription regulation is based on the action of protein-activators that enhance the activity of RNA polymerase. In *E. coli*, this positive regulation plays an important role in the activation of transcription units with relatively weak promoters, which themselves are poorly bound to polymerase. Attaching an activator protein to a specific DNA sequence located close to the promoter facilitates the "landing" of RNA polymerase, which ultimately leads to an increase in the probability of transcription.

In the above examples, the information programs that control transcription are carried out by maker proteins-regulators, which result in changes in the state of the promoter of the corresponding gene, by interacting directly with it, or with adjacent nucleotide sequences. Such programs modify the instructive gene information to work out the necessary changes in the CPS. In other words, we can see dynamic gene instruction. At any given time for reading and subsequent reproduction, only that part of it is available that is necessary and sufficient for the further production of makers executing programs needed for the given state of the cell. In addition, the management of the transcription program allows updating the rate of execution of various programs.

Let us continue examination of the task program for the manufacture of a simple maker – transport RNA, tRNA.

The final programs are executed during the post-transcriptional processing of the resulting primary transcript.

For all types of RNA, including tRNA, the primary transcript is subjected to the so-called splicing. In eukaryotic cells, in the gene segment of DNA between sections containing direct information about the subsequent synthesis (called exons) there are sections called introns in which such information is absent. Naturally, the primary transcript of RNA contains both exons and introns. The splicing program, which performs a special complex maker – the spliceosome, serves to remove the intron sequences. As a result of this program, the transcript is cut along the boundaries of the exon and intron sites, after which the exons are connected to each other, and the remaining intron sites degrade in the core.

Also, in post-transcriptional processing, the makers of RNase P and RNase D run programs to remove the 5′-leader nucleotide sequence and the 3′-terminal sequence.

The formation of the 3′-terminus of tRNA with the CCA sequence necessary for binding to the amino acid transported by the tRNA is carried out with the assistance of the maker of the 3′-exonuclease.

At the end of maturation, eukaryotic tRNAs must be transferred to the cytoplasm, where they participate in the program of protein biosynthesis – translation. The tRNA transport program is carried out along a Ran-dependent pathway with the participation of a transport maker exportin t, which recognizes the characteristic secondary and tertiary structure of mature tRNA. Presumably, exportin 5 may be an auxiliary protein maker capable of carrying tRNA through nuclear pores along with exportin t.

Let us proceed to the consideration of the aforementioned long task program for the creation of a ribosome, an integrated maker of the most important program for the creation of proteins –translation.

The first stage of this program is the independent parallel execution of transcription programs with the decisive participation of the makers of three types of RNA polymerases I, II and III, in order to obtain rRNA, template mRNA and transport tRNA. As already mentioned, rRNAs are directly included in the ribosome, mRNA is an informational instruction for the production of ribosome-forming proteins – translation, and tRNA is the maker of the assembly of the peptide chain of the translation program, which assists the main maker – ribosome.

The main features of composite transcriptional programs and subsequent processing, leading to the creation of RNAs for further use, are similar for all three types of these molecules and have been already discussed above.

The following sequentially executed component programs are programs of translation – the creation of peptide chains of all proteins that make up the ribosome, which includes rRNA and a number of specific proteins that form its structure.

Let us examine the translation program.

The main translation maker is the ribosome, which collects peptide chains of all proteins, including those that are part of it. In the process of translation, the ribosome reads three adjacent nucleotides in an mRNA molecule, which are called the codons and they encode a specific amino acid in the protein chain. Next, the ribosome captures that tRNA, which is connected to the amino acid corresponding to the codon, and attaches it to the growing peptide chain, forming a peptide bond.

Each ribosome consists of two subunits, the large and the small, which together form a complex.

The small subunit retains mRNA and tRNA, while the large one catalyzes the formation of a peptide bond. More than half the mass of the ribosome is rRNA, which plays a key role in the catalytic activity of the ribosome.

Protein biosynthesis of ribosomes begins with the stage of translation initiation. This part of the program is performed by a compound maker, formed by a small 30S sub particle of the ribosome and specific initiator tRNA.

Initiator tRNA supplies the amino acid with which begins the synthesis of the polypeptide chain.

The small subunit of the ribosome with the initiator tRNA attached to it finds, on the mRNA molecule, among all the AUG codons found here, one particular codon (start-codon) corresponding to the beginning of the creation of the peptide chain. As soon as this happens, several initiation factors that previously associated with the small subunit are separated from it, freeing up space for the large subunit of the ribosome to attach to it in order to perform the main phase of the translation program – elongation that creates the polypeptide chain corresponding to the messenger RNA.

The main elongation makers are the ribosome composed of both subunits and transport RNA.

There are three different communication sites in the ribosome: one for mRNA and two for tRNA.

Of the last two, one site holds the tRNA molecule attached to the growing end of the polypeptide chain (therefore, it is called the peptidyl-tRNA-binding site or the P-site), and the second serves to keep the newly arrived tRNA molecule loaded with an amino acid; called aminoacyl-tRNA-binding site or A-site.

The tRNA molecule is firmly attached to both sites only if its anticodon is paired with its complementary codon mRNA.

Elongation program includes three stages.

At the first stage, the aminoacyl-tRNA molecule binds to the free portion of the ribosome adjacent to the occupied P-site. Binding is carried out by pairing the nucleotides of the anticodon with three mRNA nucleotides located in the A-region.

At the second stage, the carboxyl end of the polypeptide chain is separated in the P-region from the tRNA molecule and forms a peptide bond with the amino acid that is attached to the tRNA molecule in the A-region. A peptidyl transferase maker catalyzes this reaction.

At the third stage, a new peptidyl-tRNA is transferred to the P-site of the ribosome, while the ribosome moves along the mRNA molecule by exactly three nucleotides, exposing the next codon for its corresponding tRNA.

This stage requires energy and it is driven by a series of conformational changes induced in one of the components of the ribosome by hydrolysis of the associated guanosine-5′-triphosphate (GTP) molecule.

All cells have a set of transport RNAs, each of which can carry only one of the 20 amino acids used in protein synthesis.

The connection of this tRNA with the corresponding amino acid is performed by special maker – aminoacyl-tRNA synthetase.

Each amino acid has its own specific synthetase (there are 20 such synthetases).

The last stage of translation is termination.

Of the 64 possible mRNA codons, three, namely, UAA, UAG and UGA, are termination or stop codons and they stop the translation process.

Protein makers, called release factors, are directly associated with any stop codon that has reached the A-region of the ribosome.

This binding alters the activity of the peptidyl transferase maker.

The maker with altered activity than adds not an amino acid, but a water molecule to the peptidyl-tRNA.

As a result, the carboxyl end of the growing polypeptide chain is separated from the tRNA molecule.

And since the growing polypeptide is retained on the ribosome only through its connection with the tRNA molecule, the completed protein chain gets separated from the ribosome and immediately enters the cytoplasm.

After this, the ribosome releases mRNA and breaks down into two subunits.

A number of information programs carries out the concentration control of the most important maker – ribosome, and, consequently, regulate the intensity of the execution of all programs for the production of peptide chains of protein makers and components. These programs regulate the intensity of the implementation of various compound programs that make up the task ribosomal program.

The only two component programs that we have not covered are the posttranslational formation of protein components and the program for assembling a ribosome from RNA molecules and proteins. Those two programs complete the task program that creates a complex ribosome maker.

In the process of translation program, growing polypeptide chains acquire a highly specific spatial structure, which is formed after the completion of biosynthesis. The program of a polypeptide chain formation into a spatial structure is called folding. As a result of folding in water solutions, the free energy of a water-soluble polypeptide decreases, the hydrophobic amino acid residues are packaged predominantly inside the molecule and the hydrophilic residues are located on the surface of the protein globule.

Proper folding of polypeptide chains of certain proteins in eukaryotic cells is provided by specific protein makers called chaperones, which are necessary for the effective formation of the tertiary structure of polypeptide chains of other proteins, but they are not part of the final protein structure. After exiting the ribosome, the newly synthesized proteins in order to function properly should fit into stable three-dimensional structures and remain there throughout the functional life of the cell. Maintaining quality control of the protein structure and stacking of multiprotein complexes is carried out by maker chaperones. Chaperones bind to the hydrophobic sites of improperly packed proteins and help them curl up and achieve a stable native structure.

The assembly program of the ribosome begins when each ribosomal protein recognizes its landing site on the ribosomal RNA and sits on it. So, all different types of ribosomal proteins are seated on the ribosomal RNA. At the same time, ribosomal RNA forms the nucleus of the ribosomal subunit, and most of the proteins get drawn to the periphery.

The first RNA-protein interactions begin to take place while continues the transcription of rRNA genes. Packaging takes place in cell nucleus.

Like some other programs for assembling maker complexes, for example, cytoplasmic microtubules, the ribosome assembly program is self-assembling, when the assembly makers are part of this assembly.

So, the program of ribosome assembly is the ending of the key cell task program to produce the ribosome – base maker creating peptide chains.

Provision and Energy Programs

Provision programs produce components for the subsequent synthesis of large structural cellular molecules: proteins, nucleic acids, lipids, polysaccharides and others.

These components are the so-called small organic molecules of simple sugars, fatty acids, amino acids and nucleotides.

Primarily, these molecules are synthesized by a number of bacteria and plants from the air and water environment through photosynthesis programs.

In animal cells, which are fed by the absorption of large molecules of "ready-made" proteins, nucleic acids, lipids and complex sugars, small molecules are obtained after the catabolism decomposition of large molecules by the digestive body organs. These programs are carried out under the action of enzymes secreted into the cavity of the digestive tract.

In addition, there are programs in the cell that decompose defective or short-lived cellular proteins using the ubiquitin-dependent protein proteolysis system.

For the implementation of most programs and, above all, synthesis programs, in addition to its components, energy sources are needed, which in cells are the high-energy molecules of the coenzymes ATP and NADPH. The EPs create such high-energy molecules.

In photosynthesizing organisms, these high-energy molecules are produced by programs at first light stage of the photosynthesis.

In non-photosynthetic organisms, ATP molecules formed after decomposition of complex biomolecules by catabolic programs. For cells that form a multicellular organism, the sequence of EPs includes the following steps:

1. Disintegration of large molecules in the digestive tract (outside the cell) into small molecules: proteins into amino acids, polysaccharides into simple sugars (glucose), fats into fatty acids and glycerol.
2. Transport of all three types of small molecules through the membrane into the cytosol of the cell.
3. Transformation of the main energy source of the cell – glucose – to pyruvate in the cytosol of the cell.
4. Transport of pyruvate, fatty acids and amino acids through the outer and inner membranes of the mitochondria in its matrix.
5. Disintegration of pyruvate and fatty acids in acetyl-CoA in the matrix.
6. Reproduction of citric acid from the acetyl-CoA cycle, the main result of which is the reduction of NAD to NADH, the carrier of high-energy electrons.
7. Transport of high-energy electrons in the composition of NADH to three complexes of respiratory enzymes (the first of which is NADH dehydrogenase complex), which localized on the inner mitochondrial membrane.
8. Multi-stage electron transfer along the respiratory chain to oxygen with a gradual release of energy for the process of formation and transfer of protons into the intermembrane space. Thus, creating an electrochemical proton gradient – the main electrochemical energy source in the cell.
9. Use of electrochemical proton energy by ATP synthetase is also localized on the inner mitochondrial membrane, to obtain from ADP + Pi high-energy ATP compound. This process creates the main source of chemical energy in the cell.

Executors, makers of the digestive system programs and the circulatory system of a multicellular organism, which will be discussed later in the chapter, "Multicellular Organisms", perform the first and second stages of EP.

The third stage of decomposition of glucose in order to form pyruvate is performed in the cell cytosol (the site of its execution) by component program of glycolysis, which consists of ten simple programs, each runs by separate executor, special enzyme: hexokinase, phosphoglucose isomerase, phosphofructokinase, aldolase, phosphofructokinase, dehydrogenase, phosphoglycerate, phosphoglyceromutase, enolase and pyruvate kinase.

The result of glycolysis is the conversion of one glucose molecule into two molecules of pyruvic acid (PVA) or pyruvate and the formation of two reducing equivalents in the form of the coenzyme NADH, which, in particular, are used by subsequent EPs to obtain ATP from ADP.

The compound glycolysis program is controlled through several key components of the programs performed by hexokinase, phosphofructokinase and pyruvate kinase.

At the same time, hexokinase is inhibited by the type of negative feedback by the product of the program, glucose-6-phosphate, which allosterically binds to the enzyme and change its activity. Fructose 2,6-biphosphate allosterically activates phosphofructokinase. And pyruvate kinase (L-type) is repressed by ATP and acetyl-CoA and activated by fructose-1,6-biphosphate.

The fourth stage in the delivery of pyruvate and fatty acids – components for the subsequent decomposition programs in the new relevant site – the mitochondrial matrix is carried out by appropriate transport programs.

The fifth and sixth stages include further decomposition programs, performed in the matrix of mitochondria. First, maker pyruvate dehydrogenase complex decomposes pyruvate in acetyl CoA and then acetyl-CoA is disintegrated to CO_2 and H_2O during the Krebs cycle, by the makers: citrate synthase, aconitase, NAD-dependent isocitrate dehydrogenase, α-ketoglutarate dehydrogenase complex, succinyl-CoA synthetase, succinate dehydrogenase, fumarate hydratase and NAD-dependent malate dehydrogenase.

The Krebs cycle is regulated “by the mechanism of negative feedback”, the cycle is actively working in the presence of a large number of substrates, and in case of excess reaction products, it gets stalled. Regulation is carried out with the help of hormones. These hormones are: insulin and adrenaline. Glucagon stimulates glucose synthesis and inhibits Krebs cycle reactions. Acetyl-CoA also plays another important role in cycle autoregulation as a powerful activator of pyruvate carboxylase.

At the seventh stage, the resulting NADH, the carrier of high-energy electrons, transmits these electrons to the first of three complexes of respiratory enzymes localized on the inner mitochondrial membrane, the NADH-dehydrogenase complex maker.

At the eighth stage, programs of multistage electron transfer along the chain of respiratory enzymes to oxygen with their gradual release of energy for the processes of formation and transfer of protons into the intermembrane space. NADH-dehydrogenase complex accepts electrons from NADH and transfers them to ubiquinone – a small fat-soluble molecule that transfers electrons to the second complex of respiratory enzymes – complex bc1, from which electrons through cytochrome c (a small peripheral membrane protein) are transferred to the cytochrome oxidase complex and then get transported to oxygen.

The successive loss of electron energy in each of the three mentioned complexes is used to carry out programs for the formation, transfer of protons into the intermembrane space of mitochondria, and thus create an electrochemical proton gradient – the main electrochemical energy source in the cell.

Finally, at the last ninth stage, the maker ATP-synthetase enzyme converts one form of energy into another, carrying out a program to synthesize a high-energy ATP molecule from ADP and Pi in the mitochondrial matrix during the reaction using the energy of the proton current in the matrix.

During the execution of this program, the ATP-synthetase enzyme recognizes among its molecules the molecules of ADP and phosphate, which it captures and binds freely to them ("free" state). The next change in the shape of the enzyme molecules press them to each other ("close" state), which leads to the formation of ATP. After that, the enzyme releases ATP than binds the following molecules ADP and phosphate and the cycle of ATP production gets repeated.

The so-called respiratory control regulates the intensity of the sixth through ninth key EPs. At the same time, in the passive state, the cell does not consume ATP and, accordingly, ADP is practically absent in mitochondria. As a result, ATP-synthetase is unable to use the proton gradient on the inner mitochondrial membrane. This in turn inhibits electron transfer in the respiratory chain, as a result of which NADH cannot be re-oxidized to NAD^+. The resulting high $NADH/NAD^+$ ratio slows down the Krebs cycle. And, on the contrary, activating the work of the cell increases the consumption of ATP, which contributes to an increase in the concentration of ADP and, consequently, an increase in the intensity of the implementation of all the programs from stages six to nine.

Reproducing Sites

Another group of programs provides localization of the volume, the site where the program runs, in other words it produces membranes and creates from them the so-called organelles.

Almost all of the lipids needed to build new cell membranes, including phospholipids and cholesterol, are formed as a result of running programs in the endoplasmic reticulum (ER) membrane. The main synthesizing phospholipid is phosphatidylcholine, which is formed during the execution of several successive programs of two fatty acids, glycerophosphate and choline. Each of these programs is carried out by enzymes whose active centers are facing the cytosol because all the necessary metabolites are located there. The first program is performed by acyltransferase, which adds two fatty acids to glycerophosphate to form phosphatidyl acid, a compound that is hydrophobic enough to remain in a lipid bilayer even after synthesis. The lipid bilayer increases at this stage. Subsequent programs form the "head" of the newly formed lipid molecule, thereby determining the chemical composition of the bilayer, but the growth of the membrane volume practically does not occur. All the main phospholipids of the membrane – phosphatidylcholine (PC), phosphatidylethanolamine (PE), phosphatidylserine (PS) and phosphatidyl inositol (PI) – are synthesized in this way.

Both the initial formation of phosphatidyl acid and its subsequent modifications with the formation of various types of phospholipid molecules occur on the half of the ER lipid bilayer that faces the cytosol. This could have eventually turned the lipid bilayer into a monolayer if there were no programs helping transfer the part of the newly formed phospholipid molecules to the other half of the ER bilayer. In the ER, the number of phospholipids gets aligned on both sides of the membrane in a matter of minutes.

Programs that carry out such a rapid movement across the bilayer are performed by enzymes of phospholipid translocators, which are specific for each type of the head group. Apparently, in the ER membrane there is a translocator ("flip-groove") that is capable of transferring choline-containing phospholipids (but not ethanolamine, serine or inositol-containing) from one half of the bilayer to the other. This means that PC reaches the inner surface of the bilayer much easier than PE, PS or PI. It is the programs performed by translocators that provide the necessary asymmetric arrangement of lipids in the bilayer.

The plasma membrane (Golgi apparatus membranes and lysosomes) are the parts of the membrane system associated with the ER through transport vesicles that implement programs for the delivery of proteins and lipids to organelles in these compartments.

Mitochondria and peroxisomes do not belong to this system that is why other programs import proteins and membrane lipids into them.

It was shown that special water-soluble proteins carry out such programs for the transfer of individual phospholipid molecules from one membrane to another. These proteins are called phospholipid carrier proteins (or proteins that exchange phospholipids). When executing this program, the protein "extracts" a phospholipid molecule from the membrane and detaches from it, carrying an attached lipid at the binding site. When this protein reaches another membrane, it "unloads" the bound lipid molecule into a new lipid bilayer. Carrier proteins randomly distribute phospholipids between organelles. Such random exchanges can lead to the transport of lipids from the lipid-rich membrane to the membrane depleted by them, while the phosphatidylcholine and phosphatidylserine molecules will be transferred from the ER, where they are synthesized, into the mitochondrial membrane and peroxisome. It is possible that mitochondria and peroxisomes are the only "lipid-depleted" organelles in the cytoplasm, and this "accidental" transfer is sufficient, although there could also be specific mechanisms for the transfer of phospholipids to these organelles.

Transport Programs

Now let us turn our attention to the important group of support programs – transport programs, which solve numerous problems in a cell. We will cover some of their functions.

First, they create the required concentration of component and energy molecules at program sites.

The special task of transport programs is to create gradients of the most important reaction and signal ions of hydrogen, sodium, calcium and chlorine.

In addition, special transport programs are responsible for delivering the working parts of the programs – maker in their reaction volumes – sites.

Special programs perform endo- and exo-cytosis in the cell.

All proteins are created on the ribosomes in the cytosol. They perform programs in the reaction volumes of various cellular compartments: the nucleus, the ER, the Golgi apparatus, the mitochondria, as well as in peroxisomes and lysosomes. At the same time, some proteins perform programs in the cytosol itself.

The so-called sorting signals (SSs) identify the proteins that to be sent to the different compartments. Specific receptor of transport proteins, starting a specific transport program, recognizes the SS.

The SSs are either an additional extended stretch of amino acid sequence 15–60 residues in length, which, after the transport program is completed, can be left out, or a three-dimensional structure formed by atoms of the protein surface when the molecule is coagulated.

Let us examine some examples of SS for proteins.

Proteins that must enter the ER normally carry the N-terminal signal peptide. Five to 10 hydrophobic amino acid residues form its central part. Most of these proteins are sent from the ER to the Golgi apparatus; those that have a specific four amino acid sequence at the C-terminus remain as permanent components.

Many proteins intended for mitochondria have signal peptides in which positively charged amino acid residues alternate with hydrophobic ones.

Among the proteins on their way to the nucleus, most have signal peptides formed by a cluster of positively charged amino acid residues.

Finally, some cytosol proteins have signal peptides, which covalently bind fatty acid that directs these proteins to the membranes without penetration into the ER.

There are two known transport directions for proteins after their formation on the ribosome.

In the first direction, proteins after separation from the ribosome enter the cytosol. The part of them that contains special SSs is directed by appropriate programs to the nucleus, mitochondria and peroxisomes. Another part of proteins, which does not have such special signals, remains to carry out its programs in the cytosol.

In the second direction, proteins that have special SSs, usually located at the N-end, are transferred to the ER using a special program as they are formed. As soon as the next section of the polypeptide chain of such protein is synthesized on the ribosome bound to the ER membrane, it penetrates the lipid bilayer of this membrane. Some proteins remain partially enclosed in the membrane, where they perform programs as transmembrane proteins. The rest fall into the lumen of the ER, where a part of them with specific signaling regions is retained to carry out the ER programs. Non-retained proteins are involved in further transport, which occurs with the help of bubbles that separate from one membrane to merge with another. As a result, these proteins, depending on the presence and type of SSs in them, fall into the Golgi apparatus, lysosomes, secretory vesicles or onto the cell surface in order to execute programs.

Let us consider in more detail the nuclear-cytoplasmic protein transport. The number of proteins imported into the nucleus includes histones, components of replicative systems, ribosomal proteins and transcription factors.

The inner contents of the nucleus interact with the cytoplasm of the cell through the nuclear pores – special protein complexes located in the dual nuclear membrane.

The proteins that form the nuclear pore, commonly called nucleoporins (Nup) and they form a complex symmetrical octahedral structure.

The channel in the center of the pore can vary and reach a diameter of 25 nm, allowing very large molecules and even supramolecular complexes to pass through.

The program of transport from the cytosol to the nucleus includes the following steps:

- First, the working element, maker of the initial partial program, which in this case is the α-importin protein complex with β-importin, is linked through a SS with the transported protein.
- Then, the entire trimmer approaches the Nup proteins of the nuclear membrane's pore complex.
- After that, with the participation of RanGTPase, which hydrolyzes GTP, the complex of the transported protein is transferred into the nucleus. The maker, working element of this partial penetration program is the pore complex.
- In the nucleus RanGTP interacting with β-importin performs a program to separate the transferred protein from α-importin, and β-importin combines with RanGTP and returns to the cytoplasm, where RanGAP1 performs a program to convert RanGTP to RanGDP.

The SSs are not split off after the protein enters the nucleus. This distinguishes nuclear transport from, for example, transport to the ER and other organelles. This difference may be due to several reasons:

- First, the SS can be located not at the end, but in the center of the polypeptide and take part in the formation of an active complex, for example, in the complex of RNA or DNA binding. Removal of the signal in this case may lead to loss of functional activity.
- Second, some nuclear proteins are constantly moving into the cytoplasm and re-imported into the nucleus.
- Third, during cell division, the nuclear membrane breaks up and gets re-assembled, thus the released nuclear proteins must be imported again into the newly formed nucleus.

Let us turn to the transport of proteins into the ER.

Let us consider an example of the co-translational transport of proteins in ER, in which the transfer of proteins through the ER membrane occurs as they are synthesized.

The SSs for ER usually include 5–10 predominantly hydrophobic amino acids and are located at the N-terminus of the protein. The part of SS remote from the end contains a consensus sequence recognized by a specific protease – the "signal recognition particle" (SRP), which, as a maker, starts the first simple program from the overall integrated program of the synthesized protein transfer into the ER. The composition of the SRP includes six proteins and a short RNA molecule. One SRP region binds the signal sequence, and the other binds to the ribosome, which synthesizes the protein and blocks translation. A separate SRP domain is responsible for binding to the SRP receptor on the ER membrane.

Together with the SRP, the ribosome moves to the ER and binds to the SRP receptor on the cytosolic side of the ER membrane. This complex is associated with a pore-protein translocator on the ER membrane, which is a working element of the next protein transport program. After binding to the translocator, the SRP – SRP receptor complex is separated from the ribosome, and this leads to a resumption of translation. As the translocator program is carried out, the synthesized protein enters the ER through the water channel of the translocator, which has a portal mechanism and is formed in eukaryotes by four subunits of the Sec61 complex.

After the C-terminus of the protein is separated from the ribosome and is inside the ER, a program of cutting it off from the protein is implemented by the signal peptide protease. After this, the program of protein folding into the normal conformation inside the ER is performed. With the remaining signal peptide, programs for moving it with the opened side channel of the translocator – the working element, maker of this program – into the lipid bilayer of the ER membrane and subsequent programs for its destruction, performed by the corresponding proteases, are sequentially executed.

A protein that has entered the ER remains in this organelle if it has a special "four-amino acid-retaining sequence" at the C-terminus. Some of the remaining proteins in the ER perform folding and post-translational modification programs on proteins that pass through the ER.

It should be noted that in all the considered composite sequential programs, the product resulting from the previous program serves as the object of the next program of this composite program.

An important role is played by vesicular transport, which carries out the transfer of proteins and lipids between the membranes of various organelles, as well as ensures the processes of endo- and exo-cytosis in the cell.

As an example, let us examine the process of selective cholesterol endocytosis.

This lipid is synthesized in the liver. It forms a so-called low-density lipoprotein (LDL) in a complex with other phospholipids and a protein molecule. LDL is secreted by the liver cells and then is spread throughout the entire body by the circulatory system.

Special receptors begin the sequential compound program for the transport of LDL into the cytosol and the subsequent release of cholesterol from it. These receptors of plasma membrane serve as makers in the first component program of this sequence. Diffusely located on the surface of the membrane, receptors recognize the protein component of LDL, and form a specific complex with LDL.

Then this complex moves to the zone of fringed pits and is surrounded by a membrane and plunges deep into the cytoplasm. After immersion in the cytoplasm of a fringed vesicle loaded with LDL, a rapid loss of the clathrin layer occurs, the membrane vesicles begin to merge with each other, forming an endosome – a vacuole containing absorbed LDL particles bound to receptors on the surface of the membrane. Then occurs dissociation of the LDL-receptor complex. Small vacuoles are removed from the endosome, whose membranes contain free receptors. These vesicles are recycled, incorporated into the plasma membrane, and thus, the receptors return to the cell surface. The remaining endosome merges with the primary lysosomes, forming secondary lysosomes. In them, LDL molecules are hydrolyzed to free cholesterol.

The transfer of individual molecules and ions through the membrane, in order to create the necessary concentrations of these substances in the respective compartments and organelles, implement transport programs with are carried out by special membrane transport proteins or their complexes.

Transport proteins form two groups: carrier- and channel-forming proteins.

The carrier proteins bind the molecule of the transported substance, which leads to their conformational changes and, as a result, to the transfer of this molecule through the membrane.

Channel-forming proteins form water-filled pores that penetrate the lipid bilayer. When these pores are open, the molecules of specific substances (usually inorganic ions) pass through them and, therefore, through the membrane.

All channel-forming proteins and many carrier proteins allow solutes passively pass through membranes. This process is called passive transport (or facilitated diffusion) and is carried out in accordance with the electrochemical gradient consisting of concentration and electrical gradients.

Active transport of substances against their electrochemical gradients is always carried out by proteins – carriers and proceeds with the use of external energy.

As an example of active transport let us consider a program executed by a protein complex ($Na^+ + K^+$) – ATPase, which actively pumps Na^+ outward, and K+ inside the cell against their electrochemical gradients. This is how the most important for the life of the cell ($Na^+ + K^+$) - pump works.

The binding of Na^+ and the subsequent phosphorylation of ATPase from the cytoplasm induces conformational changes in the protein, as a result of which Na^+ is transferred through the membrane and released into the extracellular space. Then, the binding of K^+ on the outer surface and the subsequent dephosphorylation of ATPase return the protein complex to its original conformation. At the same time K^+ passes through the membrane and is released into the cytoplasm. For each cycle of the pump three sodium ions and two potassium ions is transferred.

As already noted above, channel-forming proteins form pores in the membranes filled with water, and mainly serve for specific passive transport of ions; therefore, in the literature they usually appear under the name of ion channels.

Unlike performers of active transport programs, for example ($Na^+ + K^+$) – ATPase, carrying out the transfer of ions against their gradients, performers of passive programs, such as ion channels pass ions through themselves using their electrochemical gradients through the lipid bilayer of the membrane.

While this process is happening, the channels regulate the ion flow passing through them due to their selectivity with respect to various ions, depending on the size of the latter, using opening and closing mechanisms.

Channel "gates" most often open in response to specific membrane perturbations. Among them, the most well-known are changes in membrane potential (potential-dependent portal channels), mechanical stimulation (mechanically opening channels), and the binding of signal molecules (ligand-dependent portal channels).

THE MAIN OBJECTIVES OF A CELLULAR PROGRAM SYSTEM (CPS) AND ITS IMPORTANT PROGRAMS

The production programs considered by us constitute, if it can be said so, the backbone of the cell program system. Starting from the introduction of the term CPS, we will sometimes replace the term cell with it in the following presentation. This is since in the developed approach, morphologically and functionally, the cell is the result of the implementation of the CPS, which determines both its reproducible life cycles and evolutionary changes.

Before turning to the other major program groups of the CPS, in addition to the production, we consider the goals and tasks that this system pursues and solves.

The main goal can be formulated as follows – to ensure the reproduction and functioning of the cell and, accordingly, its CPS in changing environmental conditions.

To ensure this goal it is necessary to solve those problems:

- Management of the intensity and pace of implementation of all programs for their coordinated work at different stages of the cell cycle and in response to changing environmental conditions
- Development, transformation, amplification and transmission of signals in chains of control programs
- Reproduction, cell division

Control programs, signaling programs, and the program of mitosis – cell division as part of the general cell program (GCP), which implements its life cycle (which we will consider later) solve these tasks.

Control Programs

As we already know, for the execution of a program, an aggregate domain of three components: the performer – the maker, the object and energy is needed at a certain time interval in one space.

Naturally, the intensity of the program, i.e., the number of these programs running per unit of time is determined by the concentration of these components.

Program management usually comes down to controlling the intensity of their execution. Consequently, they can be controlled by changing the concentrations of the three program components.

The most effective and diverse in terms of implementation mechanisms is the variation in the concentration of program implementers, makers: proteins, RNA and complexes based on them.

Since all literature that deal with the regulation of metabolic processes in the cell use the concept of an enzyme, we will use this generally accepted term in this section, implying its expansive interpretation as a maker.

First, it should be noted that the enzyme, as a rule, can be in two states: active, in which it is able to perform its function in relation to the object (substrate), and passive, in which it cannot perform this function. When we talk about the concentration of the enzyme, we mean an active enzyme – maker. A quick and widespread way of varying the concentration of such an enzyme is to control its transition from one state to another, which is its activation or inhibition (inactivation).

The activity of the enzyme can be regulated in various ways. For example, by activating zymogen (pro-enzyme), covalent modification or retro inhibition (inhibition by the type of negative feedback),

Zymogen (pro-ferment) is an inactive precursor of the enzyme. In order for the zymogen to become an active enzyme, some part (or parts) of its polypeptide chain must be cleaved by another enzyme that performs the program for zymogen processing.

Covalent modification is called covalent attachment or cleavage of a small chemical group from an enzyme that regulates its activity. With the help of such modifications, either the completely inactive form of the enzyme becomes active, or, vice versa, the fully active enzyme is inactivated.

The performer of the program that regulates activity of the convertible enzyme (the object of the program) is a converting enzyme and it performs covalent modification of the convertible enzyme.

An amplification effect is achieved due to the fact that one molecule of converting enzyme can modify many molecules.

Often, a covalent modification is achieved by attaching a phosphate group to a specific amino acid residue of an enzyme called phosphorylation.

Phosphate comes from ATP, and enzymes called protein kinases carry out the transfer program. Subsequent removal of the phosphate group – dephosphorylation, which nullifies the effect of phosphorylation, is achieved with the help of another enzyme called phosphoprotein phosphatase.

An inhibitor in the compound metabolic program performs the retro inhibition program, in which the substrate undergoes several successive transformations and each local program gets executed by its own enzyme. Inhibition occurs if the final product of the composite program blocks one of the earlier local programs. To do this, the final product must either be structurally similar to the substrate of the blocked reaction (in other words, it acts as a competitive inhibitor), or contact any other part of the enzyme, thus regulating its activity (acts as a noncompetitive inhibitor).

One of the fastest and most effective mechanisms for regulating the activity of enzymes is the type of regulation that undergoes allosteric enzymes. Such enzymes have a catalytic center that binds to the substrate, and a regulatory, or allosteric center. The latter combines with effectors that can increase or decrease the activity of the enzyme. The binding of the effector to the allosteric center causes conformational changes in the enzyme molecule that occur at the level of the tertiary structure, as a result it changes the affinity of the enzyme to the substrate.

If the action of the effector leads to a decrease in the catalytic activity of the enzyme, this effector is called a negative or inhibitor. If the action of the effector increases the catalytic activity of the enzyme, it is called positive.

The positive effector, or activator, is most often the substrate of this regulatory enzyme.

The effectors can be end products of this program, enzyme substrates, as well as some end products of related metabolic programs.

The example of simple allosteric regulation – regulation by the "end product" of the activity of the first (key) enzyme – the performer of a consistent unbranched metabolic program. The final product as a negative effector performs a program to suppress the activity of the first enzyme, realizing the principle of negative feedback between the concentration of the final product of the metabolic program and the activity of the first enzyme – the executor of this program.

An example of this type of regulation is the inhibition of isoleucine biosynthesis. The compound program for converting L-threonine to L-isoleucine includes five consecutive programs.

The performer of the first component of the program, the enzyme L-threonine deaminase is allosteric and inhibited only by L-isoleucine.

On the surface of the L-threonine deaminase molecule, there are two types of sites: catalytic – for binding the substrate (L-threonine) and regulatory – for binding the effector (L-isoleucine).

When L-isoleucine accumulates in the cell, it binds to the allosteric center of the L-threonine deaminase enzyme, carrying out a program to suppress its activity, and the synthesis of L-isoleucine decreases to a complete cessation.

In branched metabolic programs, the activity of allosteric enzymes is more difficult to regulate, since the biosynthesis of several final products depends on the activity of the first enzyme.

The first stage of the extensive program is performed by various allosteric centers on the surface of the enzyme molecule. Each of them is associated with one of the final products that perform the function of an effector. Inhibition of the activity of this enzyme can occur in two ways:

- *Multivalent inhibition* – For this necessary binding to the allosteric centers of all end products. Each final product (effector) separately, by contacting "its" allosteric center, does not change the activity of the enzyme.
- *Cumulative* or *additive inhibition* – Addition of one final product to the enzyme partially reduces its activity, with the addition of each subsequent final product, the effect of inhibition increases.

In some branched programs, the inhibition of the first enzyme is carried out not by the end products of each of the branches, but by an intermediate product formed during the execution of a component of a program just before branching.

Its accumulation is in turn controlled by end products. This type of inhibition is called sequential.

There are extensive metabolic programs in which regulation is carried out in a way that both activation and inhibition occur simultaneously.

Thus, the regulation of the activity of the executor of the program (the enzyme in a narrower sense) is a common, effective and operational way to solve three important tactical tasks of cell functioning.

First, is the maintenance of a stable equilibrium state of a cell, characterized by a well-defined ratio of the intensity of cell programs execution at each stage of the cell cycle.

Second, is the response of the cell to changes in the conditions in which the cell functions (the change in the intensity of specific programs compared to their stationary intensity).

Finally, is a return to the stationary ratio of program intensities after the cessation of the changed conditions.

To solve the strategic tasks of restructuring the functioning of the cell widely used regulation of the synthesis of enzymes (performers of programs), for example, when changing stages of the cell cycle, while the significant restructuring of the entire system of cellular programs is in process, by significantly changing the intensity of their execution.

Such regulation can be carried out at different stages of the composite program for the synthesis of polypeptide and RNA molecules when performing particular programs:

- Transcription of gene information from a DNA molecule
- Processing of the primary transcript
- The transport of mature mRNA molecules from the nucleus to the cytoplasm, where the translation program is carried out
- Translation – assembling a peptide protein molecule according to the information contained in mRNA
- Degradation of mRNA, which determines the number of protein assemblies committed using this mRNA

In addition, regulation is also possible at the level of implementation of programs for the formation of the final "working" conformations of protein and RNA, as well as the creation of enzyme complexes.

The regulation of the transcription program is one of the most effective ways.

As a typical example of transcriptional regulation, the synthesis of the β-galactosidase enzyme in *E. coli* bacteria was analyzed when changing the growing medium containing glucose to the medium with lactose. The content of the β-galactosidase enzyme involved in the breakdown of lactose had been increased 1000 times. Simultaneously with β-galactosidase, the synthesis of two more enzymes is induced: β-galactoside permease, which provides lactose transport into the cell through the cytoplasmic membrane, and β-galactoside transacetylase, which are necessary for the utilization of lactose.

This is due to the fact that the genes responsible for their synthesis are located nearby at the chromosome, forming the so-called Lac-operon, and are triggered by one mechanism in response to the effect of the inducer – lactose. Transcription of Lac-operon genes leads to the synthesis of one common mRNA molecule.

In addition to the structural genes listed above, the Lac-operon contains a promoter-operator region. The operator is a small segment of DNA bordering the first structural gene and partially overlapping with the promoter. The operator binds the repressor, which, by blocking the binding of RNA polymerase to the promoter, interferes with the transcription of structural genes.

The repressor – an allosteric protein that has two binding sites: the operator recognizes one center; the other interacts with the effector or inductor, which in this case is a lactose derivative, allolactose. After interacting with the inductor, the repressor goes into an inactive state and disconnects from the operator, freeing the promoter. As a result, transcription of the Lac-operon's structural genes is realized.

The repressor's protein gene is located near the structural genes.

In the absence of lactose, the repressor molecule, which is active in the free state, performs a program of negative regulation of the synthesis of Lac-operon enzymes.

This molecule binds to the operator, while closing the promoter, which prevents the binding of RNA polymerase and the start of transcription of structural genes.

With an external inductor in the environment, β-galactoside permease transports lactose into the cell, and the β-galactosidase enzyme performs a program to convert it to allolactose, which acts as an internal inductor. The β-galactoside permease and β-galactosidase enzymes are also present in non-induced cells, but in amount, that is less than 0.001 from their concentrations after completed induction.

Allolactose performs a program to induce the synthesis of Lac-operon enzymes. It binds to the repressor, which in this case undergoes a conformational change that reduces its affinity for the DNA of the operator, and as a result, the repressor is detached from Lac-operon. Transcription of structural genes begins, leading to the synthesis of enzymes of lactose catabolism.

When the allolactose inductor is removed from the cell, the repressor again goes into an active free state and performs a program of negative regulation of the synthesis of Lac-operon enzymes, communicating with its operator, which leads to the termination of the synthesis of the corresponding enzymes.

Lactose operon is also subject to regulation of another – positive type.

The fact is that RNA polymerase can bind to a promoter and start a program for the transcription of structural genes only after the CAP regulatory protein (catabolite activator protein) performs its activating program and joins the promoter.

In turn, the CAP can perform its program only in the presence of cyclic adenosine monophosphate (cAMP) in a sufficiently high concentration. Thus, the presence of cAMP is the beginning of the program for the Lac-operon transcription.

The presence of glucose in the cell dramatically reduces the concentration of cAMP and, thereby, represses the synthesis of structural genes, including β-galactosidase. This phenomenon is called catabolite repression.

Glucose catabolic repression can be removed if cAMP is added to the medium. A complex of cAMP with a CAP is formed, and RNA polymerase is joining the promoter and begins the implementation of a program for the synthesis of lactose catabolism enzymes, even in the presence of glucose.

Signal Programs

For the correction and fundamental changes in the state of the program system in the cell, the most common and effective way is to regulate the activity and synthesis of the program executor, the enzyme. Executors of such regulation programs are often called effectors.

Typically, effectors are the final link in a chain program that is started, or rather activated by a signal.

Most often, the signal is just an external impact on the cell, thermal, electrical, or the most common chemical, that forces the cell to adequately rebuild its program system.

The cell perception of external signals is mainly due to the interaction of external factors with cell receptors located on the outer membrane. By performing the first program in the chain, the receptors recognize the external signal and activate the intracellular pathways of information transfer leading to the launch and regulation of various intracellular processes.

In the literature, the act of changing the activity of an enzyme or the efficiency of its synthesis is often referred to as a cellular response. In our terminology, the cellular response records the change in the efficiency of the execution of one or several programs of the system (change in the state of the CPS).

Often in the process of performing a composite, "complete" signaling program between the activation of the receptor by the signal and the cellular response, programs are performed with the participation of so-called mediators. As such, either protein complexes or relatively small molecules can serve, for example, cAMP.

With the participation of mediators, a substantial amplification of the very weak signal perceived by the receptor primary signal is usually made.

There are three main types of receptors integrated into the outer cell membrane: receptors coupled to G-proteins; receptors-ion channels and receptors associated with enzymatic activity.

Receptors associated with G-proteins (abbreviated to GPCR, from G-protein coupled receptors) are widely distributed. They transmit the primary signal to intracellular targets by performing the first program in a composite program implementing the GPCR → G-protein → effector protein cascade.

The same primary signal can initiate signal transmission through several (sometimes more than ten) different GPCRs, GPCRs are monomeric integral membrane proteins, the polypeptide chain of which crosses the cell membrane seven times.

The receptor domain responsible for interaction with the primary signal is localized on the outer side of the membrane, and that in contact with the G-protein is located on its cytoplasmic side.

G-proteins are heterotrimers that consist of three types of subunits: a, b and g, but in vivo, the last two subunits function as a single bg-complex.

The most important characteristic of G-proteins is the presence of a guanyl nucleotide-binding center on their a-subunit: GDP and GTP.

If GTP is bound to G-protein, this corresponds to its activated state (G-GTP), if GDP is present in the nucleotide-binding center, then this form (G-GDP) is not active.

In the first signal cascade program, an activated receptor catalyzes the exchange of G-protein-bound GDP for the presence in GTP medium.

This GDP/GTP-exchange on the G-protein is accompanied by the dissociation of the trimeric G-protein molecule into two functional subunits: the a-subunit containing GTP and the bg-complex.

In the next program, one of these functional subunits, which one depends on the type of signaling system, interacts with an effector protein represented by an enzyme or a cation channel. Therefore, their catalytic activity or ionic conductivity changes accordingly. After that, the following program performed by the effector protein leads to a change in the cytoplasmic concentration of the secondary messenger (or cation), which, in turn, and ultimately initiates a particular cellular response.

When the signal is transmitted in a cascade, the receptor → G-protein → effector protein, the original external signal can be amplified at many times.

Thus, one molecule of the receptor during its stay in the activated state (R*) manages to carry out the program for the translation into the activated form (G*) of many G-protein molecules. For example, in the visual cascade rhodopsin → Gt → cGMP-phosphodiesterase per molecule R* several hundreds or even thousands of molecules can be formed, which means that in the first stage of the cascade R* → G* the gain of the external signal is 10^2–10^3.

Further, amplification occurs at the stage of program execution by the effector protein, when a large number of second messenger molecules appear (disappear) in the cytoplasm. Thus, in the visual cascade of its second stage, one molecule of activated cGMP-phosphodiesterase can split up to 3000 cGMP molecules per second, which serves as a secondary messenger in photoreceptor cells.

As a result, the total gain of the cascade may exceed 10^6.

After the termination of the external stimulus, all components of the signal system should be «turned off».

At the receptor level, this is achieved, first, as a result of the dissociation of the primary signal from the complex with the GPCR, second, by receptor phosphorylation programs under the action of special protein kinases and the subsequent program of binding a special protein (e.g., b-arrestin) with a modified receptor.

G-proteins have the ability to hydrolyze the associated GTP to GDP, which ensures their self-exclusion, that is, the transition G-GTP → G-GDP.

The effector protein also goes into an inactivated state, and, as a result, the synthesis program (hydrolysis) of the second messenger or the closing of the ion channel stops.

Finally, special programs restore the initial level of the secondary messenger or cation in cytoplasm.

For example, the cAMP messenger whose cytoplasmic concentration increases when the signal is transmitted in the cascade b-adrenoreceptor → Gs-protein → adenylate cyclase then as a result of running the program by cAMP phosphodiesterase hydrolyzes to non-cyclic AMP, which does not have the properties of a secondary messenger.

Receptors-ion channels are integral membrane proteins consisting of several subunits, the polypeptide chain of which crosses the outer cell membrane several times. They act simultaneously both as ion channels and as receptors, which are able to specifically bind from their outer side primary signals that change their ionic (cationic or anionic – depending on the type of receptor) conductivity.

The primary signals for such receptors are some of the neurotransmitters responsible for synaptic transmission in electrically excitable cells.

Receptors associated with enzymatic activity, according to the mechanism of work are divided into two groups.

The first group includes receptor-enzymes, from the cytoplasmic side of which there is a catalytic site activated by the action of an external signal on the receptor. These include protein tyrosine kinases, which carry out programs for the phosphorylation of target proteins. It is also true for protein tyrosine phosphatase, realizing dephosphorylation programs. And finally, there are receptor guanylate cyclases, which carry out programs for the synthesis of a second messenger, cGMP.

The second group of receptors under consideration does not possess its own enzymatic activity. However, in the presence of an external signal, they acquire the ability to bind cytoplasmic (non-receptor) protein tyrosine kinases, which are inactive in the free state, but active in combination with the receptor and carry out a program for its phosphorylation. In turn, the phosphorylated receptor binds target proteins, which, after phosphorylation, transmit a signal further down the cascade.

The General Cellular Program and Its Most Important Component – Mitosis

The goal of the GCP is the realization of the cell cycle: from the newly released daughter (small in size, with single DNA) cells to the mother cell, which is divided into two daughter cells.

A GCP consists of program systems that solve the following cycle tasks: growing the cell size to the optimum, DNA replication, and the implementation of mitosis (the division of a cell into two subsidiaries).

Typically, the cell cycle is divided into two main phases – the M phase (division phase or mitosis) and the interphase (intermediate phase).

In the M phase of mitosis, two groups of program carries out the division of the cell into two subsidiaries.

One of them separates the chromosomal material doubled by this time, separates it into opposite cell poles and creates a dual-core cell by successively carrying out prophase, prometaphase, metaphase, anaphase and telophase programs.

The main executors of this group of programs are microtubules of two pole centrosomal centers and performers of programs that disassemble the nuclear structure at the time of dilution of daughter chromatids at the poles and the subsequent assembly of two nuclei around each group of poles of chromatids.

The second group of programs, called cytokinesis, divides a two-nuclear cell into two between its poles, which creates two single-core cells.

The main performer of the second program group is a located on the membrane contractile ring consisting of actin and myosin filaments. It pulls the membrane inward, dividing the cell into two.

In the interphase program, the cell is grown to the size at which the onset of mitosis is possible, and the cell is prepared for division, doubling the centrosomal center and chromosomes (DNA replication) and preparing for the manufacture of performers of mitosis programs: microtubules of the centrosome center and contractile ring, partially realizing their production.

The interphase consists of three successive stages: phases G1, S and G2.

In the G1 phase, the intensive biosynthetic activity of the cell resumes, sharply slowed down during mitosis. At the same time, the following operations are carried out: re-growth of the cell and its compartments to normal volumes; replenishment of material resources – objects of metabolic programs; increasing the concentration of the working elements of programs – proteins, RNA and complexes based on them, and also, replenishing the energy reserves of the cell.

All this is a consequence of the implementation of the considered groups of programs: procurement, energy, transport, maker and membrane.

In addition, the Gl phase programs prepare the beginning of the doubling program of the centrosomal center, realizing the divergence of two centrioles by several microns, and DNA replication, by activating the synthesis of RNA and proteins necessary for this, as well as controlling the structure of the DNA and realizing, if necessary, the repair program.

In the S-phase, two composite programs perform the main task of chromosome doubling, one of them reduplicates of the DNA molecule itself, and the other carries out the packaging of both DNA molecules of the old, maternal and new, daughter into a pair of identical chromosomes, sometimes called sister chromatids. Sister chromatids are closely approximated and connected only in the region of the centromere. The histones, necessary for the construction of new chromatin, are synthesized by specific programs at high speed only in the S-phase.

A program for the reparation of copied DNA accompanies the replication program, if necessary.

In addition, in the S-phase, the implementation of programs for doubling the centrosomal center begins, by forming daughter centrioles near each old centriole at right angles to it.

In the G2 phase, the cell prepares for division.

Executed all the programs that grow the cell as a whole and all its organelles to the size, quantity and mass necessary for the cell to enter the M phase.

EPs intensively accumulate energy to perform high-energy mitotic programs.

Programs that directly prepare phase M synthesize microtubule proteins, which will form a spindle during mitosis.

The implementation of programs for the growth of daughter centrioles and the formation of two centrosomal centers is completed.

As in the previous two phases of interphase, programs are carried out to detect and after detection to repair DNA damage.

In addition, in the G2 phase, the cell controls the accuracy of the DNA reduplication that has occurred and corrects the detected failures.

The cell prepared for division enters the phase M.

All of its organelles and the cell as a whole are of the necessary size. The number of chromosomes is doubled. The required amount of energy has been accumulated. The centrosomal centers were prepared – the performers of the main preparatory program of mitosis, the separation of chromosomes between the poles. Conditions have been created for the rapid creation of the performer of the division program itself – the contractile ring.

Several basic programs are performed during the course of each stage of mitosis.

At the first stage, called prophase, a program for chromatin condensation is performed, as a result, all chromosomes consisting of two sister chromatids get interconnected in the region of the centromere and become clearly visible in the microscope. At the end of the prophase, the cytoplasmic microtubules, which are part of the interphase cytoskeleton, disintegrate and begins the program for the assembly of a bipolar spindle from microtubules and related proteins.

The next stage of prometaphase begins with a program that implements the disintegration of the nuclear membrane into small membrane vesicles, as a result the microtubule of the spindle, which was outside the nucleus penetrates into the nuclear region.

At the same time, a program for the formation of special protein complexes called kinetochores at each chromosome centromere, for communication with microtubules is being carried out.

As the kinetochores are formed, a program of spindle microtubule attachment to them begins to be performed. The microtubules attaching to kinetochores are now called kinetochore microtubules. The rest of the spindle microtubules are called pole, and those that lie outside the spindle are called astral. The primary attachment of microtubules to kinetochores does not have a polar orientation, in other words, the connection of any kinetochore with microtubules of both poles is equally probable. However, the situation when the kinetochores of different sister chromatids are associated with microtubules of different poles turns out to be more stable. Therefore, after some time of program flow, different kinetochores of a chromatids pair are connected to microtubules of different poles.

This leads to the fact that at the metaphase stage, microtubules, as the program runs, bring each chromosome into the equatorial plane halfway between the poles of the spindle, forming a so-called metaphase plate. This happens due to the equality of forces with which microtubules pull each of the sister chromatids to opposite poles.

The anaphase phase begins with a program that breaks the centromeric link between the sister chromatids of each chromosome, after which its two chromatids slowly begin to diverge toward their respective poles under the influence of the microtubules pulling forces.

When this happens, programs for changing the length of microtubules are executed: kinetochore microtubules shorten as the chromosomes approach the poles, and the polar microtubules lengthen, and the spindle poles move further away from each other.

At the onset of telophase, as a result of ongoing anaphase programs, the divided daughter chromatids approach the poles, the kinetochoral microtubules disappear, and the polar microtubules continue to lengthen

The main programs in telophase form a nuclear structure around each group of daughter chromatids, thus creating, at the time before dividing, a two-core cell.

The division of a dual-core cell created in telophase into two single-core cells is implemented by a contractible ring's program at the cytokinesis stage.

The preparatory program for assembling the reducible ring on the cell membrane in the plane perpendicular to the bipolar spindle from the actin and myosin filaments begins in anaphase.

The assembled reducible ring begins to contract and retract the membrane inward toward the axis of the spindle, forming a division groove, which gradually deepens until it reaches the remains of the spindle located between the nuclei, where for some time there will be a bridge between the two daughter cells, called the residual body.

After a short period, the residual body is destroyed, and the cells are completely separated.

CELL CYCLE MANAGEMENT

As already noted, the goal of the GCP is the realization of the cell cycle: from the newly isolated daughter (small in size, with single DNA) cells to the mother cell, which is divided into two daughter cells.

GCP consists of program systems with phases G1, S, G2 and M, which solve the following cycle problems.

In the G1 phase, the intensive biosynthetic activity of the cell is resumed; the following are carried out: the cell and its compartments are re-grown to normal volumes; material resources are replenished (objects of metabolic programs); concentration of the working elements of programs is increased (proteins, RNA and complexes based on them), and the energy reserves of the cell are replenished.

In the S-phase, the problem of chromosome doubling and DNA replication is solved, and the program of doubling the centrosomal center begins.

In the G2 phase, the cell prepares to divide.

All programs that grow the cell as a whole and all its organelles to the size and mass necessary for the cell to enter the M phase are executed.

EPs intensively accumulate energy to perform high-energy mitotic programs.

Programs that directly prepare phase M synthesize microtubule proteins, which will form a spindle during mitosis.

The implementation of programs for the growth of daughter centrioles and the formation of two centrosomal centers are completing.

Finally, in the M phase, a program of mitosis is carried out – one cell divides into two.

In the cell cycle, the existence of so-called "checkpoints", or restriction points, is postulated, the passage of which is possible only in the case of the normal completion of the previous stages and the absence of breakdowns.

We know of at least four such points: a point in G1, a point in S, a point in G2, and a point in mitosis.

In a reconciliation test in G1, the basic requirement for a cell entering the S-phase, DNA intactness, is verified, since replication of damaged DNA will lead to the transmission of genetic abnormalities to offspring.

Therefore, cells that have undergone mutagenic effects which causing DNA breaks (UV and g irradiation, alkylating compounds, etc.) stop at G1 and do not enter the S-phase.

The process of stopping in G1 is observed not only after DNA-damaging effects, but also in other conditions, including disruptions in the number of chromosomes – if the previous cell cycle is incomplete because of mitosis (chromosomal divergence), and when chromosomes are not properly segregated during mitosis, leading to the formation of micronuclei, as well as the destruction of microtubules, which can later cause disruption of mitosis.

The calibration point in the S-phase controls the accuracy of DNA replication.

The calibration point in the G2 phase controls DNA damage and other abnormalities that were missed during the passage of previous calibration points or were obtained in subsequent stages of the cell cycle. In addition, in the G2 phase, the completeness of DNA replication is detected, and cells with not fully replicated DNA are not included in mitosis.

The restriction checkpoint of the spindle assembly delays the M phase until all kinetochores are attached to microtubules.

The essential parts of cell cycle (CC) program management scenarios are:

- Regulation of the intensity of metabolism, leading to cell growth and energy storage in it (phases G1 and G2).

- Management of division processes (phases S and M).
- Control of passage of restriction points.

Different phases of the cell cycle and transitions between them are controlled by programs performed by protein makers, which are both positive and negative regulators.

The positive regulators include, first, the complexes of cyclins and cyclin-dependent kinases and the transcription factors of the E2F family.

The negative ones are pRB, p53 proteins and p15, p16 – inhibitors of cyclin-kinase complexes.

In all these cases, the variation of the quantity and concentration of makers is widely used to manage the relevant programs. This is achieved, in particular, by transcription factors of the E2F family's programs, affecting the transcription intensity of the corresponding genes that play a key role in cell division programs.

In general, E2F factors largely determine the implementation of the general program of the cell cycle.

Inactivation of E2F occurs when the inhibitory pRB protein is attached to it. Activation of E2F occurs in the case of phosphorylation of pRB, after which it leaves the complex with E2F.

For example, in the middle and at the end of the G1 phase, phosphorylation of the pRB protein and subsequent activation of E2F are carried out by the cyclin kinase complexes CycD/Cdk2 (or CycD/Cdk4, CycD/Cdk6).

Control of the activated state of E2F is one of the main functions of cyclin-dependent kinases (Cdk, CDK).

Different types of CDKs, referred to as CDK1–CDK6 in the order of their discovery and associated with cyclins of different variants, are involved in the regulation of different phases of the cell cycle.

For example, CDK1 is associated with cyclins A and B and is involved in the G2-M transition, CDK2 can bind with cyclins A, E, D2 and D3 and is one of the main kinases that regulate the transition of G1-S and passing through S-period, CDK4 and CDK6 are involved in the regulation of the G1-S transition.

CDKs are able to perform their functions in the cell cycle only after their interaction with the corresponding cyclins and the implementation of post-translational modifications under the action of CAK and other similar proteins regulating the cell cycle.

Regulation of the activity of CDK is carried out at the expense of directional changes in the level of certain cyclins in certain phases of the cell cycle.

The formation of this complex becomes possible after the cyclin reaches a critical concentration. In response to a decrease in the intracellular concentration of a specific cyclin occurs a reversible inactivation of the corresponding CDK. More than one cyclin activates some CDKs. In this case, a group of cyclins, as if transferring protein kinases to each other, keeps them activated for a long time. Such CDK activation waves occur during the G1 and S-phases of the cell cycle.

A number of protein makers (including p15 and p16) inhibit the activity of cyclin kinase complexes.

Initiation of the division program is carried out by proliferative signals, which are very diverse, depend on the cell type and are formed either by special programs of other cells of the body, or by programs inside the dividing cell itself. Such signals can be various growth factors, interleukins and hormones. Programs of certain receptors recognize proliferative signals on the cell surface, which activate the composite programs of intracellular signaling. At the first stage of these programs, activation of transcription factors (c-Ets, c-Jun, c-Myc, c-Myb, B-Myb), encoded by the genes of the early response, is carried out. Further, these transcription factors carry out programs for the activation of the synthesis of cyclins and cyclin-dependent kinases, which are encoded by the late response genes and are the executors of the main programs controlling all stages of cell division.

Thus, programs that control the cell cycle form information chains, starting with registration of detection of initiating signals and ending with the launch of programs for the manufacture of makers producing controlled programs of the cell division itself and preparation for it and/or activation of these makers.

In the middle of these chains are intelligence information makers – deciders (from the English word decide), which determine, choose the upcoming actions, the steps of generating programs depending on the signals that come from the initiators of proliferation, for example, size, level of energy resources, etc.

Typically, such makers, deciders, among which already mentioned complexes of cyclins and cyclin-dependent kinases, transcription factors of the E2F family, proteins pRB, p53, p21 and others, form information program networks (IPNs). IPN arise through the exchange of deciders by internal signals, which often lead to their phosphorylation (or dephosphorylation), as a result of which they change the level of their activity and, accordingly, give out or do not their control information signal. The combination of such signals in the IPN is the governing decision in this situation, determined by external (initiating) and internal (characterizing the state of the cell) factors.

The p53 protein and the local information network formed around it play a key role in the implementation of cell cycle checkpoints.

So an important role in the delay of cells in the G1 phase is played by the p53 – regulated p21 protein, which blocks the activity of CDK2 and CDK4, thereby preventing the phosphorylation of pRB, triggering the transcription of genes such as cyclin K, hCDC4, p53RFP and the complex of genes responsible for the synthesis DNA.

P53-induced genes BTG2 (suppresses cyclin E1) and MCG10 also contribute to cell delay in the G1 phase.

The p53 protein can also cause a delay during the S-phase of DNA synthesis, for which, apparently, is responsible one of its alternative isoforms involved in the induction of 14-3-3σ and p21 protein.

Damage and malfunctioning in the S-phase causes ATR-dependent induction of Chk1 kinase, which modifies p53. However, the main function of full-length p53 in the S-phase is not to delay the cell cycle, but to stimulate DNA repair.

Delay in the G2/M phase is important to prevent segregation of damaged and under-replicated chromosomes (chromosome segregation is the process of longitudinal chromosome splitting into chromatids [daughter chromosomes]). Cessation at this point occurs due to the suppression of the activity of the Cdc2-cyclin B complex, which is promoted at once by several p53-induced genes – GADD45, BTG2, GTSE-1, REPRIMO, HZF and MCG10.

OTHER CPS PROGRAMS

These are designed to help cells survive:

- *Homeostasis* – Maintaining intracellular homeostasis, i.e., stability of the basic conditions necessary for the implementation of all programs: temperature, osmotic pressure, acidity level, etc., in sites, cell compartments.
- *Antistochastic* – Counteracting the growth and accumulation of stochastic defects in the cell, carrying out the repair of erroneously paired nucleotides during DNA replication and proteolysis of "outdated" proteins.
- *Protective* – Fighting against parasites, especially with viruses, and other biological enemies of the cell, as well as with heat shock effect and other adverse effects of the external environment that are critical for the existence of the cell.
- *Evolutionary* – Implementation of evolutionary adaptation, for example, by variable splicing of primary RNA and recombination of DNA.

EXTERNAL PROGRAMS OF UNICELLULAR ORGANISMS

So far, we have considered, basically, only internal programs of cells, i.e., those that run in compartments surrounded by an outer membrane.

In contrast to internal programs, external programs are performed outside the outer membrane, either directly on the outer side of this membrane, or in the close proximity to it.

The few single-cell external programs include, first, endocytosis, exocytosis and chemotaxis. All these programs provide cell metabolism with the external environment, which, in particular, creates the material and energetic base for executing all of the cell programs, both internal and external, without exception.

The most important external unicellular program is chemotaxis – the movement of bacteria in the external environment toward attractants (mainly nutrients) and from repellents (e.g., toxins and substances that destroy the membrane).

We will analyze the movement of bacteria using the example of *E. coli*.

This bacterium moves in a liquid medium due to the rotation of its flagella. Each bacterium can have six or more flagella distributed randomly on the cell surface. Each flagellum consists of spirally laid molecules of one protein – flagellin. Due to flagella twisting into a spiral, the direction of their rotation significantly affects the nature of the movement of the bacteria.

When rotated counterclockwise, all flagella gather in a common bundle, and the bacterium moves in one direction.

However, when rotated clockwise, the bundle crumbles into individual flagella, and the bacterium begins to randomly tumble in one place.

In the absence of a chemotactic signal, periods of rectilinear movement are interrupted by short periods of tumbling that randomly change the direction of movement. In the presence of an attractant, tumbling is partially suppressed when the bacterium randomly moves toward its greater concentration. This leads to a gradual approach of the bacterium to the attractant.

The general program that solves the problem of chemotaxis, i.e., the movement of the bacterium toward the attractant and from the repellent, consists of two main subprograms: the program of rotation of the flagella causing the movement of the bacterium, and the program controlling this movement. Let us consider these subprograms in more detail.

The flagella rotation program (FRP), in turn, consists of four particular programs.

First, it is a program of movement of a flagellum in an environment in which its inner end creates a circle, thereby forming the necessary hydrodynamic force required for movement of a bacterium.

A hook attached to the inner end of the flagellum with its outer end performs the following motion conversion or transmission program. In this case, the inner end of the hook is connected to the rotor of the cell electric motor and rotates coaxially with it, and the outer one, connected to the flagellum, as already indicated, moves along a circular trajectory.

The rotor of the electric motor, together with the stator covering the rotor, performs a program for converting the energy of the proton gradient into rotational motion.

And, finally, a separate program creates a proton gradient on different sides of the cell membrane at the location of the electric motor and flagellum.

The program of controlling the rotation of flagella (PCRF) switches the direction of this rotation for the time corresponding to the gradient of attractant or repellent.

PCRF includes a signal chain triggered by an external chemical signal attractant or repellent, and ending with an effector that controls the direction and duration of rotation of the molecular electric motor of the flagellum.

Chemotactic stimuli are detected by receptors known as methylated chemotaxis proteins (methyl-accepting chemotaxis proteins [MCPs]), which are membrane sensors.

The interaction between the receptors and the flagellum switch is carried out by the four proteins – CheA, CheY, CheW and CheZ.

CheA while interacting with the signal domains of the MCP forms a secondary signal in the chain.

CheA is a protein kinase that phosphorylates itself in the presence of ATP, and after that quickly transfers the phosphate group to CheY, phosphorylating it.

CheY protein, being an effector, binds to FliG proteins of the rotor's motor-switching complex of an electric motor and, depending on the degree of phosphorylation, causes the flagellum to rotate in a specific direction.

In the absence of an attractant, the concentration of phosphorylated CheY is maintained at a sufficient level that not only favors the rotation of the flagellum clockwise but also causes the chaotic movement of the bacterium.

The binding of the attractant to the sensory domain of the receptor induces a conformational change in its signaling domain, which using the CheW adapter, (the CheA protein) suppresses the auto kinase activity of the one that is bonded to it. As a result, the concentration of phosphorylated CheY is decreased, and the flagella of bacteria rotate counterclockwise for a long time.

Thus, the main chain of PCRF consists of three successive programs executed by the chemotaxis receptor MCP, protein kinase CheA and effector CheY.

Programs adapting the sensitivity of the receptor to changing, in particular the growing concentration of attractant, are performed by the CheR proteins, which methylate MCP and CheB dimethylating MCP. The methylation program has the opposite effect to the binding of the attractant. Methylation is stimulated by binding of the attractant to the receptor, the degree of methylation increases with increasing concentration of the attractant and eventually neutralizes the effect of attractant's binding. However, at any given moment of increase in the concentration of attractant, there is a part of the active receptor that is uncompensated by methylation, which forms the secondary signal, causing first the phosphorylation of CheA and then CheY. When an attractant is removed, the receptor is demethylated by the program performed by CheB.

A FULL PROGRAM SYSTEM OF UNICELLULAR ORGANISMS: MANAGEMENT PRINCIPLES

By such a system, we mean the totality of all programs that run on a given single-cell phenotype and create this phenotype.

The purpose of the system of programs (SPS) is to implement such a large number of cycles of self-reproduction and reproduction of the phenotype that will ensure adaptive changes with the help of evolutionary change of the genotype.

To achieve these goals, the SP must solve the following tasks:

- Maintain the number of groups and programs in them at a necessary and sufficient level. At any given time to have the ability to reproduce and vary the intensity for any program of the system, providing for this the necessary level of material and energy resources.
- Maintain the conditions necessary for the functioning of programs, primarily providing cell compartmentation and realization of homeostasis in these compartments
- Provide the necessary pace of reproduction of cell division cycles
- Ensure anti-entropic renewal, purification, correction and treatment of the structural elements of the programs, primarily performers, and compartments.
- Ensure adjustment of SP and phenotype to changing conditions.
- Implement the management of SP programs, ensuring the maintenance of an adequate for this phenotype state of SP.

The previously considered groups of programs, which are categorized according to their results, provide the solution of all these problems.

1. **Maker programs**. First, these are the programs that create makers performers of all programs. The most important of them are the programs of transcription and translation.

   As we already know, program performers include proteins and RNA, as well as complexes of them such as homogeneous, for example, microtubules formed by self-assembly from a single protein, and heterogeneous, for example, ribosomes collected from dozens of different proteins and RNA.

2. **Resource, procurement programs**. The next group of programs performs the procurement of resources – the necessary material (molecular, atomic) and energy, in the form of high-energy molecules and the electrochemical potential for the implementation of all SP programs. In addition to intracellular programs, this group includes such external programs as endocytosis and chemotaxis in its attractant part.
3. **Transport programs** provide transfer of makers, objects and energy molecules of programs to their sites, as well as the operation of signal chains and energy pumps.
4. **Compartment programs** form all the cell membranes, including the outer and organelle, as well as the elements of its cytoskeleton.
5. **Homeostasis programs** monitor the stability of the basic conditions necessary for the execution of all programs in the cell: temperature, osmotic pressure, acidity level and others.
6. **Signal programs** detect signals, which are specific molecules or physical effects, for example, radiation, temperature changes, etc., as well as their, signals, conversion, amplification and transfer to performers of controlled programs. The signals characterize the state of both the external environment and the internal parameters and characteristics of the cell.
7. **Control programs** carry out a change in the intensity of the implementation of individual programs in accordance with the situation determined by the combination of both external and internal signals. For the most part, control programs modulate the concentration of makers in controlled programs, which are proteins, RNA and their complexes in the active state. Such modulation is often carried out either by activating/inactivating the maker, or by regulating the programs of its synthesis at various stages, primarily of transcription and translation.

   The result of control programs is the reorganization of SP programs system.
8. **Group of programs for reproduction**, generation of the hereditary cell includes DNA replication and the implementation of mitosis – the division of a cell into two.
9. **Anti-entropic programs (AEPs)** are programs that counteract the growth, accumulation of the entropy of the phenotype and the genotype of the organism, caused by stochastic processes of degradation and aging.

   This name implies, in particular, programs that limit the lifetime of "aging", i.e., accumulating stochastic defects, performers of all programs consisting of proteins and RNA. This is primarily a program of ubiquitin-dependent protein degradation.

   Also considered to be AEP are lysosomal and exocytosis programs that withdraw from the cell slags and debris.

   Of course, the most important AEP programs are reparation ones that restore degrading DNA.
10. **Protective programs** include immune programs to combat parasites, especially viruses. They also comprised of repellent chemotaxis programs, programs that fight against heat shock and others that protect the unicellular

organism from the adverse effects of the external environment and biological enemies.

11. **Evolutionary programs**, such as, for example, variable splicing of primary RNA and recombination of DNA, carry out trial variations of program makers, peptides and RNA, which, together with subsequent selection, lead to the evolutionary adaptation of the organism.

The 11 program groups constitute the necessary and sufficient programmatic set, allowing a single-celled organism to accomplish all the tasks for achieving the above objectives.

The first seven groups are common, their results are necessary for the implementation of all programs.

The eighth specific group implements the culminating stage in the life of the cell – its division.

The last three: anti-entropic, protective and evolutionary, perform key functions to ensure maximum extension of existence of the SP are also not common.

We call the programs of the last four groups from the eighth to the eleventh special.

Of the four groups of special programs, three are reproduction, anti-entropy and protective programs, they are all tactical, i.e., the result of their implementation is manifested during the lifetime of the organism. Whereas, the evolutionary programs forming the fourth group can be considered strategic, since, they affect the phenotype in time comparable to the lifetime of this phenotype.

Changes in the state of the program system occur both on regular cyclic signals and on stochastic, irregular signals, which can be both internal and external.

The internal signals of the first type are the signals that determine the phases of the division cycle, in them large role play cyclins. As a result of the launch of the corresponding control programs by cyclins, a change occurs in the intensity of the execution of the general programs of the seven groups and the start-up stops of the special reproduction programs of the eighth group.

A typical example of external cyclic signals is periodic changes in environmental parameters such as temperature, light, etc., caused by astronomical causes, namely, the rotation of the earth around its axis and its rotation around the sun. These signals, as a rule, cause a change in the intensity of the execution of general programs.

Regular changes in the state of the SP in response to regular signals are mandatory. When we spoke earlier about the steady state of the SP, we meant a state that corresponds to the phase of normal, regular changes.

In the future we will replace the term "stationary state of the SP" with the term "regular state of the SP".

Irregular signals may be stochastic changes in external conditions, attacks by enemies of the body, for example, viruses, or local fluctuations in the intensity of the execution of various programs. All these signals trigger signal chains, the effectors of which directly or through control programs change the intensity of the execution of general programs and, if necessary, launch special programs to counteract those non-standard factors that gave rise to the corresponding signals.

After the action of the abnormal factor ends and, therefore, the signal triggered by them is terminated, the SP returns to its normal state.

The intensity of program execution is determined by the concentrations of the activated performer, program subject, and energy in the form of special energy molecules, high-energy parts of the object molecule, or ion gradient at the program execution site.

A common factor for all programs is energy, the resource of which is limited in any given organism. Therefore, in the case of SP energy consumption close to the limit, the need arises for restructuring the SP toward higher priority programs, i.e., those that are executed in the forced mode, as compared to the lower priority ones.

For example, special programs for dividing the seventh group in the S and M phases have a higher priority than general programs, except for those that directly provide the execution of special programs.

To maintain the regular state of SP, the intensity of primarily general programs should be maintained at a constant stationary level, which is achieved in particular by using feedback between the product of the program and the activation of its maker. These types of activation/inactivation signals can be called local.

On the other hand, it is possible to speak of global signals causing a significant restructuring of both the general and special parts of the SP, for example, in the event of the influence of proliferating factors.

As was described when reviewing the cell cycle, control of the SP is generally carried out by an IPN of signal chains, in which deciders play a key role. They are the proteins that activate or inactivate the makers of various programs depending on the set of external or internal signals.

The most prominent example of such a decider is the p53 protein. Scientists have described a number of conditions that can activate p53:

- Depletion of nucleotides
- Cytoskeleton disorders (actin fiber polymerization disorders, microtubule depolymerization)
- Ribosome biogenesis's disturbances
- State of hypoxia and ischemia
- State of hyperoxia
- The absence or excess of certain growth factors or cytokines
- Violations of cell adhesion and contacts
- Defective integrins
- Appearance of polyploid cells
- Formation of micronuclei
- Destruction of the chromosomal spindle, and others

After receiving signals from the corresponding receptors, p53 generates control signals and sends them either adjacent to the network deciders or directly to the makers of controlled programs to restore the cell's state.

The unique distinction of p53 from other deciders, for example, cyclin-kinase complexes, pRB protein and others, is its multimodality both in "input" and "output".

As for the input modalities, there is an extensive network of factors in the cell, the interaction of which determines the degree of ubiquitination and the rate of destruction of the p53 protein, and, consequently, its cellular concentration, for example, the ubiquitin ligase Mdm2, regulated by the p14ARF protein.

In addition to the systems that regulate p53 concentration levels, modifications in the p53 molecule itself play an important role in changing its activity, which affect not only the amount of protein, but also the quality characteristics of its activities.

Of great importance are covalent modifications of the p53 protein molecule, expressed in phosphorylation, acetylation, methylation, as well as the introduction of ubiquitin residues and ubiquitin-like proteins SUMO and NEDD8.

Many proteins, interacting with p53, make covalent modifications in its structure.

These proteins include more than 30 different protein kinases, several protein phosphatases, several ubiquitin ligases and proteins that regulate the interaction of p53 with E3 ligases, deubiquitinating proteins, proteins that bind SUMO and NEDD8, several methylases, a number of acetyltransferases and deacetylating enzymes.

At the output, p53 regulates expression of the:

- Genes involved in cell cycle control
- Genes involved in DNA repair processes
- Genes regulating angiogenesis and other tissue reactions
- Genes involved in the induction of cell death
- Antioxidant genes
- Genes affecting metabolism

We can distinguish two qualitatively different modes of functioning of p53.

Within physiologically tolerable deviations, with moderate stresses, physiological loads, dietary disorders, minor inflammatory processes, etc., the activity of p53 at its relatively low, basic level is aimed at ensuring homeostasis, maintaining an adequate level of reparation, mobilization of optimal sources of energy resources, switching processes of biosynthesis and ensuring the protection of the genome from mutagenic effects from excess oxygen radicals.

When the physiologically tolerable level of damage is exceeded, the task of p53 will be to get rid of genetically dangerous defective cells, which is achieved either by activating apoptosis or through the terminal release of cells from the division process, which is a form of genetic death.

In summary, we can call the information protein maker p53 the general decider in the IPN that controls the cell program system.

EVOLUTION OF UNICELLULAR ORGANISMS

We have considered the full set of programs that are necessary and sufficient for the evolving existence of cells and unicellular organisms for at least the entire observed time of evolution, i.e., several billion years.

The absence of at least one program group in the system should lead to irreversible destruction of the entire system of cellular programs and the death of the organism.

That is why, walking along the evolutionary chain to its beginning, we run into the earliest prokaryotic organisms that already have the entire set of programs we have considered.

It becomes obvious that intermediate forms between inanimate matter and the cell prototype which already has the full program package, are non-viable.

In particular, the existence of a cell prototype without a membrane is impossible. The expression "at a certain stage of evolution, a cell has acquired a membrane" has no more sense than, for example, the following: "at a certain stage of evolution, a frog has acquired skin".

To the famous expression of Frazzetta in his "Complex Adaptations in Developing Populations" is that the evolutionary problem gradually improve the machine without interrupting its operation, it should be added that the machine must work from the very first step of evolution. To do this, from the very beginning it must have the necessary and sufficient set of programs discussed above. It should be noted here that the machine is improved from the inside, by "germinating" and securing adequate, beneficial changes to the program system.

Finally, the impossibility of stochastic generation of the living from the non-living follows from a more general statement that the process cannot spontaneously transform into a program, because, as we have seen, any program can exist only within a program system consisting of a necessary and sufficient set of separate programs.

PROGRAM-INFORMATION CELL STRUCTURE

In the previous sections of this chapter, we have already covered examples of cellular programs performed to accomplish various, basic tasks. Of course, not all programs were considered, and in some cases, the level of consideration was quite superficial. But nevertheless, according to the author, the proposed material allows us to consider the declarative formulation of the main differences between the regulated live program and the stochastic process of the inanimate part of nature, made at the beginning of the chapter, reasonable and confirmed. Familiarity with a much broader information that is not directly included in this book because of the limitations of volume, further convinces the author of the validity of the proposed concept.

The degree of this conviction is reflected in the proposed slogan: *in a living cell there are no reactions and events that are not program actions performed by makers or agents stimulated by them: ions, atoms and molecules.*

The author invites all the readers to detect "non-maker" events in a living, "regular" cell and inform him about them.

It will not be superfluous to try at the end of the chapter to formulate some of the revealed elements of the program structure and the principles of its organization. Special attention will be paid to the role of the IS in its more expanded, in comparison with traditional, representation.

Let us start by listing the conditions of the program.

First, there must be three necessary components: the program executor as the maker, the object to perform actions of this particular maker and the site that is the meeting place for the maker and object in order to execute the program.

Those important conditions must also be present on the site:

- Energy transferred to the maker to perform an action in the form of energy particles or an electrochemical potential
- Required homeostasis parameters, first of all, temperature, pressure and chemical composition of cytosol

The main types of maker's actions (program steps) are mechanical, such as dividing an object into parts, assembling an object and moving an object, and other are chemical, leading to a change in its chemical state, for example, by phosphorylation, acetylation, methylation, etc. It should be noted that most of the actions are both chemical and mechanical.

A maker can be a "transforming" one, performing a state transformation of the object, or it can be a "structural" one that forms the structure of another composite maker (e.g., a ribosome and microtubule) or a membrane and cytoskeleton.

In some cases, in addition to the makers, their agents can execute programs, ions, atoms, and molecules generated and/or transported to the site by the corresponding makers. Such agents include, for example, ions of calcium, potassium, sodium, etc.

Usually the maker has several functional domains.

First, an "object domain" into which "its" object is captured for the program subsequent execution.

In addition, there are IDs of several types.

These include areas of makers into which informational agents fall – molecules or ions that activate or inactivate a given maker, examples of which are the traditionally considered coenzymes.

In the case of a negative feedback regulating loop this information agent can be a transformed object – the result of the maker action.

The domains can also recognize the characteristic portions of "their" object, for example, the gene promoter or the address region of the transported protein.

Makers consisting of many proteins and ribonucleic acids (polymerases, ribosomes, etc.) have a complex structure and additional domains, such as one for linking additional makers with them to perform a program, for example, the ribosome has a domain for landing the auxiliary maker – transport RNA with another amino acid.

We call a maker performing one action – simple. A maker performing in relation to an object more than one action will be called complex. An example of a simple maker is a protein that performs a catabolism program, and an example of the complex one is a ribosome.

Usual algorithm for the behavior of a simple maker includes the activation phase of the information agent, test contact with the object, recognition of the object and the formation of a working contact maker-object, then implementation of the program maker's action on the object and, at last, disconnection of the maker and the modified object. The energy required to perform a program action comes to the program site, for example, from the surrounding cytosol or brought with the object.

The object that is modified according to the type of action is the result of the program. As a rule, the result of the action of a program executed by one maker will be the object for the program of the next maker.

Programs executed by one maker will be called simple programs.

In the case of a simple maker such programs are single-active or one-step programs. Examples of such programs are separate catabolism programs in the Krebs cycle.

If a simple one-maker program is performed by a complex maker, then such a program is multistep, for example, the program executed by the ribosome.

Programs that combine two or more simple programs that lead to a single result are called composite programs.

Composite programs are performed by more than one maker.

Examples of composite programs are the Krebs cycle consisting of single-step simple catabolic programs and the assembly of a peptide chain performed by a complex maker – ribosome and simple makers – transport RNA.

The result of the composite program in some cases is the task result.

The task results are determined by the task classification of the programs that make up the cell's program system. They are the manufacture of makers, the reproduction of membrane sites, the production of energy molecules and the creation of electrochemical potentials, the destruction of "old" makers, etc.

For example, result of a compound program executed in a site – the cell nucleus, including transcription programs and subsequent processing is the solution of the problem creating a ready-to-work maker – transport RNA.

However, in many cases, solving a problem requires the execution of several composite programs in different sites. For example, the task of obtaining a protein maker is solved by performing two composite programs – transcription in the nucleus and translation in the cytosol of the cell.

A set of programs leading to a task result is called a task program, an example of which can be the sequentially executed programs of transcription, RNA processing, translation, folding of peptide chains, and finally, assembling ribosomes from RNA and proteins, the result of which is the production of a complex maker.

In the implementation of task programs, transport programs play an important role, carrying out, in particular, the transfer of makers and objects to sites where subsequent composite programs are executed.

The most important criterion of the program is its effectiveness, i.e., the number of realizations of its result per unit time. The relative partial performance of a separate program in a cell forms the state of the CPS. Thus, it determines the functioning of the cell as a single organism.

Therefore, a key control area of individual cellular programs is the regulation of their effectiveness.

Effectiveness is determined by two parameters: the speed of the individual program and the frequency of its reproduction.

Usually the nominal speed of program execution varies within a small range.

In this regard, the main control mechanisms are focused on adjusting the frequency of program reproduction.

The frequency of program reproduction is determined primarily by the concentrations of the makers, their objects and energy particles at the site. It is on these concentrations that the probability of convergence of the main participants of the program actions depends.

The determinant is the management of the concentration of makers on the program execution site.

The required concentration of makers on the site is provided two main programs, programs that produce makers and transport ready makers to their "working" site.

The most important of them is the first one and in it is a composite transcription program.

Not surprisingly, the abundance of mechanisms, regulating the frequency of this program in the cell, use combinations of various regulatory proteins that interact with the promoter part of the transcribed gene.

The transcriptional control is using transcription factors to bring the program cell system into a state that is adequate to either internal signals, which correspond to the next phases of the cell cycle, or external signals that cause, for example, cell differentiation.

In general, the control of the most important factor – the concentration of the active, current maker on the site is carried out in a cell at different hierarchical levels of programs that provide this factor. Including at the level of simple programs, the pace of their execution, as mentioned above, can be regulated by a feedback circuit when the product of this program is an activator and/or inactivator signal for its maker.

In more detail, the variants of control programs have been considered in one of the previous sections of this chapter.

In the previous section of this chapter, we introduced the classification of groups of cellular programs in accordance with the tasks they solve.

Here we will divide all the programs and the makers who carry them out into two basic groups – effector and information.

The group of effector programs includes all production programs providing the cell with makers, initial small-sized objects, energy, sites and homeostasis and ensuring the fight against stochastic degradation and enemies, evolutionary variations.

Information programs provide control, synchronization, cooperation of all cellular programs, producing, transforming and using information in the form of signals and instructions.

We have already noted above that the algorithm of the work of the maker includes information stages: activation in response to the signal of its information agent and recognition of its object according to the instructions, which the latter contains in the form of a specific chemical and geometric configuration.

Essentially, the key definition of an ordered program, as opposed to, the antipode of the stochastic process, in all elements of this definition implies the operation of some mechanism ensuring the implementation of these terms – reproducible, regulated and predictable.

The author reached a conclusion that such a mechanism, or rather mechanisms, is contained in IS that has a broader meaning than the traditional one.

We will try to present the basic elements and principles of such a system, relying solely on the material of this chapter, which describes the cell's program system.

First, we will look at several examples of the manifestation of information mechanisms in conditions necessary for the execution of a program: the regulated nature of the maker's actions, the choice of a specific object, the regulation of the place and time for the execution of program actions.

Reproduction by the maker of the regulated action is provided by the design of this maker, produced according to the gene instruction, which is the spatial configuration of certain signs. The assembly makers recognize this instruction using their ID and then assemble a new maker by the scenarios discussed earlier.

Performing an action on a specific object is provided by the chemical-spatial mark of the object, which is recognized by the maker performing the action using his ID, which most often coincides with the object domain.

The specific site (location) of the action of the maker is given by the address instruction, located, as a rule, in the tail part of the peptide chain of this maker. The transport maker, reading this instruction using its ID, delivers the current maker to the specified site.

Finally, one of the mechanisms for regulating the execution time of a program action is the activation of the maker by his information agent, for example, coenzyme. Immediately after that the maker performs his action. In this case, the maker recognizes its information agent, which falls into a separate ID.

What is common in all these examples? First is the presence of information marks (IMs), affecting the maker, and, secondly, the presence of IDs of this maker, reacting selectively to these marks.

The only possible interaction between mark and ID, after which a specific action takes place, is a key element of program regulation.

When performing a transcription program, two parts of the gene instruction are used: the first corresponds to the promoter – the DNA segment preceding the gene itself and the second to the gene itself. In both cases, the instructions are marks that have a strictly defined spatial configuration of the "letters" of the genetic code's alphabet – four nitrogenous bases – adenine, guanine, cytosine and thymine. Domains of the maker, transcriptional complex formed with polymerase, have the corresponding response structure.

The chemical spatial marking of an object, determined by its external structure, consists of characteristic elements – marks, which form the configuration that is aligned with the ID of the maker that works with this object.

The targeted instruction of the transported protein, called the SS in the literature, is either an additional extended portion of the amino acid sequence 15–60 residues in length that can be cut off after the transport program is completed, or a three-dimensional structure formed by the atoms of the protein surface when the molecule coagulates. The domain structure of the transport maker is configured for a specific SS.

At last, the configuration of the ID of the activated/inactivated maker selects from a broad range its own information agent.

So, the base of the regulation program is in:

- Recognition by the ID of the maker with corresponding mark and
- The subsequent strictly defined reaction of the maker to this recognition

Thus, the necessary and sufficient number of basic information elements is an obligatory information component of the program system – the IDs of all makers without exception and the IMs that clearly correspond to them.

Three types of marks can be distinguished – instruction mark, identifier mark and signal mark, examples of which are:

- Mark-instruction – DNA nucleotide sequence, according to which the RNA polymerase performs a transcription program
- Mark-identifier – the structural-chemical part of an object, by which the maker recognizes its object, or the promoter part of DNA, by which the RNA polymerase recognizes its gene
- Mark-signal is an ion, an atom, a molecule, which, when it enters the domain of the corresponding maker, activates the last.

Let us consider some actions and programs in which marks of different types take a passive part (as objects) or active (as makers).

Among all the programs performed by the makers with respect to passive instructional marks, we will cover:

- **Replication** – copying complete DNA instructions for the cell – the heiress
- **Transcription** – conversion of partial, gene information from a DNA code to an RNA code
- **Reparation** – restoration, repair of damaged elements of DNA instructions
- **Recombination** – the introduction of evolutionary variation in the DNA instruction.
- Finally, the central program is the **translation** – reading the RNA instructions and making peptide chains – blanks of the most common makers: proteins and complexes based on them.

Bearing in mind the results of these programs, we will refer to them as effector ones.

It is also necessary to introduce into this group of programs (carried out with the participation of instructive marks) the most important programs for the modification of promoter regions of DNA instructions using regulatory proteins, which determine the on-line state of the CPS.

Simultaneously with the execution of the first two programs occurs significant increase in "instances" of instructional marks – their reproduction.

The object's identifier mark also plays a passive role in relation to the maker, which also, like the instructional mark, does not change the state of the maker, but only facilitates the execution of an action on the object.

Unlike the first two types of marks, a mark signal changes the state of the maker and therefore can be considered an active element of interaction with the maker, which plays the passive role of an object in this interaction. As we have indicated, the role of the maker, i.e., active participant of the program action, in some cases, can play not makers, but agents, which are most often products of other makers.

After exposure to the inactivating mark-signal, the maker "falls asleep", turns off from the program activity. Whereas, when activated this maker is turned on and performs the prescribed action as far as contact with its object.

If the "awakened" maker is informational, then it performs an informational action, a common form of which is the launch, the generation of a secondary mark signal. In this case, such a maker participates in the formation of a signal chain.

Often at the beginning of the signal chain an information maker is located. It is called the receptor and we can find it on the external side of the cell membrane. The ID of such a maker is tuned to a specific, external mark-signal. Such a primary signal may come from other cells of a common multicellular organism, for example, it may be a proliferative factor. For unicellular organisms, the primary receptor signals are environmental agents, for example, attractant and repellent molecules in external chemotaxis programs.

The end of the information chain is the ID of the effector maker, whose action is the response of the cell to the primary signal. The cases of the direct reaction of the effector maker to the primary signal, i.e., implementations of a single-link signal chain in a cell are rare. Usually in the chains there are links formed by secondary signals, or as they are often called "messengers".

The functions that these internal links perform are very important.

In particular, it is the amplification of the signal effect by multiplying the number of secondary signal marks.

And, very importantly, intelligent makers (IMs) – deciders (as we called them earlier) in the internal links perform the function of deciding whether or not to affect the effector maker at the end of the signal chain, taking into account additional signals that affect the activity of such an IM. At the same time, the activity of IM can be determined not only by a change in its state, but also its concentration. In the latter case, the signal that controls the activity of IM, does this, for example, by regulating the transcription of the IM gene. The IM can perform logical operations, for example, logical addition or multiplication with two signals, one of which came to it from the previous link of the signal chain, and the other led to a change in its activity. The signal that changes the activity of IM, as well as, the primary receptor signals can be both internal and external, i.e., displays the situation both inside and outside the cell.

Thanks to the connections between deciders, the signal chains form networks for managing program groups and the entire cell's program system, for example, the cell cycle, as was illustrated in the previous section. The goal of the tasks solved by such an IPN is the development of a system of control effector signals that bring the program cell system to a state adequate to those receptor signals that characterize situations inside and outside the cell.

A unique example of such a master decider is the p53 protein. The main feature of its uniqueness is the multi-modality in both "input" and "output". P53 is essentially a member of the family of signal chains that also forms its own information network. A huge amount of research on this protein, which continues at the present time, allows P.M. Chumakov (the author of a review on this protein) to state: "The role of p53 in the body can be compared to the role of a conductor in the orchestra, its functions monitor the implementation of programs developed by evolution, patterns of cell behavior in a variety of conditions".

In the Section "Cell Cycle Management" of this chapter, were listed examples of other deciders involved in the formation of cellular information networks.

Now it becomes clear that there is a specific mechanism of harmonization and regulation of a huge number of local and seemingly autonomous cellular programs. This is the cumulative work of signal control programs that are performed by IMs – deciders that organized into information networks.

Summary of the Role of the Cell Information System

The regulation of program actions and programs consisting of them, as well as the coordinated implementation of these programs requires:

- Reproduction of makers specialized for this object
- Identification of objects and addresses of their delivery
- Optimizing the frequency, intensity of individual programs and their groups

The key action of information programs is recognizing "one's own" mark by the corresponding ID of the effector or information maker and the subsequent reaction of this maker to his mark by either activating/inactivating it or performing its inherent action, solve these tasks.

Instructive marks of DNA and RNA contain information on which components of all makers are collected: proteins and RNA.

Identification marks are mainly used for identification by the maker of their object.

Mark-signals form signaling program chains that often unite into information networks, solving the tasks of synchronized program management for an adequate cell response to changes in the internal and external situation.

The Possible Use of Commonly Used Informational Terms and Concepts When Describing Cellular Programs

Concepts of a Bit and a Pattern

For example, the copied or readable part of the DNA molecule is a pattern. The letter of the alphabet of the genetic code, one of the nitrogen bases is the elementary mark.

We can say that the elementary mark is the minimum part of the pattern, perceived as information by the corresponding ID.

This definition of the elementary instructional mark seems to be analogous to the definition of a bit in a computer's storage device. The pattern in this case is byte or machine word.

Memory and Storage Devices

Chromosomes containing cell instructional information, chained DNA molecules, is a long-term memory device. The array of information in this device is structured on several levels, according to individual chromosomes and gene regions. Each readable gene block has an address in the form of a promoter region, where it can be recognized by the recognizing domain of a complex maker based on RNA polymerase. DNA information is stored for a huge number of lifetimes of generations of a unicellular organism, undergoing only periodic replication during cell division, repair of defective sites and recombination transformations of evolutionary variability.

Matrix RNA – one of the possible informational patterns of the transformed gene DNA information is already an analog of the operational memory, which is used for reading and subsequent production of a peptide chain by the makers – ribosome and transport RNA. The lifetime of such information is significantly less than the duration of the cell cycle.

Information Operations

The basic information operation in a cell is the recognition of the mark by the maker domain. At the point in time when the chemical-spatial structures of the mark and domain coincide, changes that occurred in the state of the maker lead to changes in its activity or the generation of a secondary signal.

IMs such as deciders and program information networks formed with their participation, carry out a significant part of logical operations. The possibilities of activating/inactivating deciders with different signal marks in combination with the multiplicity of their IDs and output multimodality (e.g., when the p53 decider forms signal marks for several information and effector makers) allow not only to perform all elementary operations such as negation, conjunction, disjunction, but also to solve with them quite complex logical tasks.

If we continue the consideration of the example of the operation of p53, then the information operations can also include convolutions and sweeps of patterns from the mark-signals. So, the pattern of signals of the input situation is folded in a protein into a single reactionary internal signal. This signal triggers, turns on the output situation of the action, which is a set of a pattern of secondary information and effector signals, and in this case, we can speak about the sweep operation.

Spatio-Temporal Transformations

A significant part of the program contains a transformation of the spatial, geometric structure formed by instructional marks into a sequential time series of signal marks. For example, the geometric pattern of the gene is converted into a sequence of signals that determine the movement of the RNA polymerase maker along the DNA molecule and the assembly of the RNA chain. In this case, the actual assembly of this chain is the inverse transformation of the time signal series into the spatial structure of RNA.

Time, Space and Causality in the Program-Information Structure of the Cell

First, it should be noted that the program is

- sequential row
- following each other
- in a specific course of action
- directed to the result

This allows with a high degree of accuracy to call the program a temporary sequence, taking into account, among other things, the unidirectionality and vector of this notion.

Thus, the notion of time is inseparably linked with the definition of a program.

As for the notion of space, as already noted, the information base for reproducing a program system is a DNA instruction that has a strictly controlled spatial structure, which is translated into program time sequences.

Therefore, the concept of space, as well as time, is at the core of interconnected initial fundamental notions of life.

In the previous material in this chapter, "space-time manifestations" were mentioned. These include maintaining and regulating the pace of program execution, cyclic sequence of the phases of the cell cycle, reproducing cell compartments, and an external chemotaxis program as well as many others.

Causality

The source and, therefore, the unequivocal cause of all actions with objects are the corresponding makers and/or their agents. Thus, causality is one of the basic properties of the life program structure along with space and time.

Those regulatory, sequencing concepts are not part of the inanimate stochastic world. They are used by representatives of the living world, including humans, when performing external programs and contact with objects and phenomena of the inanimate part of nature.

Definition of the Information System

The IS serves to reproduce, interconnect and synchronize programs, as well as to realize the phases of the cell cycle and to create adequate responses to changes in the internal and external situations.

The basis of the IS is made up of marks and maker domains that recognize their.

The essence of IS are operations and programs that are performed with the participation of IM and correspond to the conceptual and structural notions adopted in traditional computer science

We also include the space-time apparatus and the concept of causality in the cellular IS.

Mathematical representations of measures and sets, manifested in the materials of the chapter, we also include into the cell IS.

The openness of the IP definition consists in the generalization of all means reproducibility, regulation, and adequate response of cell programs.

2 Multicellular Organisms

SOCIALIZATION IS THE GENERAL PRINCIPLE OF STRUCTURING A SPECIES COMMUNITY

The most important property of life is the presence of a species community, within which each individual and organism, realizing its own program system (PS), on the one hand, fixes a certain average online state of this species, and on the other hand modifies the PS in accordance with changing environmental conditions, adapting to them. Next the mechanism of intraspecific selection of promising adaptive modifications of the PS and their subsequent fixing should work. For this, there should be an intraspecific information system of communication and "exchange of experience" between individual organisms. Moreover, such a system should, in one form or another, already exist in the simplest unicellular with ability of nonsexual reproduction.

If we call such a universal intraspecific information system – information sociality – using a well-known and even popular term, we believe that we will avoid large exaggerations.

Further, we can formulate the following position: informational sociality (in our extended understanding) is inherent in any species. The common elements of the informational mechanism of sociality in the organism are: first, the source of the external signal "I'm yours (species)" and, second, the receptor of the same external signal from a neighboring individual, the organism, which can be summarized by the term communication. Thus, we come to the second position: communication is a property of any species. The exchange of the simplest message "I'm yours", i.e., a statement about one's affiliation is an example of an elementary communication information act, after which, for example, an effector act of bringing together single cells is possible.

The property of communication, intraspecific interaction of unicellular organisms, was the basis for the emergence of multicellular organisms (MOs) at the next evolutionary stage.

TRANSITION FROM UNICELLULAR TO MULTICELLULAR

Compared to the diversity of internal cell programs that we have previously considered, and which form the basis of the system of program (SP) of unicellular organisms, their external programs (EPs) are not numerous. These, above all, include programs of movement in the external environment such as in water with the help of flagella and on a solid surface with the help of filopodia. It is also the program of endocytosis – the absorption of the resources necessary for the functioning of the internal procurement and energy programs of the cell from the external environment. There are also the signaling programs of detection of external, primarily chemical, signals. The combination of these programs makes it possible to implement the main comprehensive program for the supply of internal programs with resources from the external environment. So, the chemotaxis program, which is a set of motion and detection programs useful for the cell molecules in the environment, transports the cell to the medium with the maximum concentration of resource molecules, where the endocytosis program effectively absorbs them.

Exocytosis should also be attributed to EPs – programs for removing unnecessary harmful particles and molecules from a cell, as well as isolating signaling molecules. Endocytosis products can serve as external signals for other unicellular organisms.

The chemotaxis and endocytosis programs allow unicellular organisms to locate each other and, if necessary, move closer or farther away from each other while showing the above-mentioned social behavior.

AGGREGATE ORGANISMS

The use of these single-cell capabilities in evolution has led to the emergence of such precursors of multicellular ones as an aggregate complex of single cells that are formed in adverse environmental conditions, such as, *Mycetozoa*.

Mycetozoa is an aggregate organism formed by unicellular myxamoeba that feed on bacteria and yeast, and divide every few hours under optimal conditions.

In response to starvation, a program for the production and extraction of cyclic adenosine monophosphate (cAMP) molecules is carried out in myxamoeba. This signal triggers single-cell chemotaxis convergence programs.

At the same time, starvation leads to the launch of many other programs, including external ones that are necessary for the formation and functioning of the aggregate organism.

These are the programs providing adhesion to each other when myxamoebas are gathering together. In the first hours of the fasting, the cell adhesion program by a Ca^{2+}-dependent mechanism involving an adhesive molecule, called contact site B, is used. Then, to perform the adhesion program, the Ca^{2+}-independent mechanism with the molecule contact site A is used.

The assembled and stuck together myxamoeba forms a migrant *Mycetozoa* – an aggregate organism of the first phase. To do this, in particular, the cells carry out new programs for the formation and secretion of mucus that envelops the entire resulting body. In this phase, *Mycetozoa*, as a single organism, performs an EP of movement and migration, in which many cells constitute an aggregate.

After the migration, the second phase of the existence of the *Mycetozoa* begins. Different cells start to perform their specific programs by internal signals. Thus, the cells located in front slide down to form a sporulation stem, while cells that were previously located behind move upward to form spores. As a result, the slime mold acquires the appearance of a plant consisting of a stem with a fruiting body filled with spores, which after scattering and entering a favorable medium turn into new myxamoebas.

On the example of an aggregate organism of a *Mycetozoa*, it is possible to single out some of the most important transformations of the SPs that occur during the formation and subsequent functioning of MOs.

This is, above all, the simultaneous appearance of "pre-aggregate" programs in individual cells: unification, adhesion and other.

It is also a collective, in many respects, synchronous execution of "aggregate" programs by individual cells, controlled by common signals, for example, during the movement of a plasmodium or the formation of a plant-like form.

In the end, we can see the emergence of several types of PSs that correspond to the so-called differentiation of cells that have different functions in an aggregate; for example, forming a stem and spores in the second phase of the existence of *Mycetozoa*.

BASIC SOCIAL PRINCIPLES OF CELLULAR INTEGRATION INTO A SINGLE STRUCTURE

As we can see, social behavior is also manifested in aggregate organisms in the implementation of specific cells, specific programs for the entire organism as a whole, which is also true for the cells with other specific programs, for example, nutrition, reproduction, and movement. Such cells, by spending their resources on the implementation of specific social programs receive a gain in the form of a guaranteed supply of resources, effective movement and reproduction. Unlike single-cell organisms, the social behavior of cells that make up aggregate organisms turns out to be resource-beneficial and evolutionarily promising.

For example, cell adhesion programs in a single aggregate organism are performed first by an adhesive molecule called contact site B and controlled by a Ca^{2+} signal, and then by contact site molecule A. Contact sites A and B were identified as integral proteins of the plasma membrane glycoproteins.

The cells containing social makers are called differentiated, and the programs for the formation of social makers in them are called differentiation. Thus, a cell as part of an aggregate organism has its own specific social makers, which in effect execute EPs with respect to this cell itself, but internal within the whole organism. "Standard" makers carry out internal cell programs, often referred to as "housekeeping", having much in common with all types of organism cells.

The social principle of uniting individual cells is seen in "full-fledged" MOs.

PROGRAM SCHEME OF A MULTICELLULAR ORGANISM (MO)

SOCIAL MULTICELLULAR ORGANISM STRUCTURE

In case of an MO, sociality is the interaction of its constituent cells and, accordingly, the systems of a cell programs (SCPs) with the aim of forming and functioning the system of MO programs (SMOPs), which is a higher order with respect to its components – SCP.

There are two groups among the tasks solved by the SMOP. The first is the task of maintaining the vital activity of all components of the MO cells, above all supplying them with the necessary food components and oxygen. The second group includes tasks whose solution is essential for MO as a single organism, for example, the tasks of reproduction and immune protection against pathogenic microorganisms (PMs).

In order to accomplish these tasks, the MO cells of its components form a system of social makers. This system has several levels.

First, these are already mentioned social makers inside the differentiated cells, which we will call the elementary makers (EMs) in the future. EMs run their programs at the same time with the rest of the intracellular housekeeping makers and compete with them for cellular resources.

Further, these are cellular makers (CM) – differentiated cells that make up the MO and perform social programs, for example, cells of the gastric mucosa secrete components of the gastric juice (GJ) for the program of grinding and dissolving food components, the muscle cells of the same stomach provide the program for mixing these components. In essence, a CM is all MO cells that are at different stages of differentiation and perform certain social programs in accordance with the control signals – nerve and/or humoral, which they perceive through membrane receptors.

In the organs of MO, cell makers (CMs) form collective tissue makers (TMs) necessary for effective synchronous execution of programs, for example, collective makers of the mucous and muscular tissues of the gastric shells.

The most important component of the social structure of MO is the organismic system of resource distribution of molecular components of nutrition and oxygen across all cells, which is the blood supply system (BSS). In addition to performing the task of transporting resources to all cells, the BSS solves the problem of collecting and integrating the products of CMs performing social programs. Thus, there is a constant exchange of social cellular products for resources through the BSS.

Another essential systems of control of all cellular social makers are the MO's nervous and the endocrine one's. Each CM and their organ and systemic associations transmit to the hierarchical sections of the nervous system (NS) signals about the state of their social programs and then receive control signals from the information makers of these departments.

MO Tasks and Program Systems That Solve Them

So, the main goals for which the cells are ready to pay with their resources in this case are the same as in aggregate organisms. It is primarily an efficient supply of cells with building materials and energy. To achieve these goals, the basic tasks of nutrition and respiration are solved in the MO.

An important task of the MO, which is solved by the immune system, is to maintain an adequate state (biological individuality) of the organism by detecting and subsequently destroying the foreign agents, which can be a pathogen, foreign body, toxic substance or a degenerated cell of the organism itself.

Finally, the most crucial task for the evolutionary survival and selection is reproducing of the MO through breeding.

The listed MO tasks are basic and are solved by the corresponding basic nutrition, respiration, reproduction and immune program (IP).

The corresponding programs will be called complete task programs, and the constituent programs included in them that solve local, particular problems will be called local task programs.

Internal and External Programs

Similar to the situation in single-celled organisms, when considering a SMOP, it is advisable to distinguish two groups of programs: internal (implemented within the organism itself) and external (performed outside the body, in contact with the external environment of existence).

In order to perform both internal and EPs in the body, there are internal and external makers localized in the external and internal organs and body systems.

Each of the basic tasks of the body – nutrition and reproduction – is solved when performing a complex of external and internal programs. In this case, we can talk about a complex task program implemented by a complex maker consisting of external and internal makers.

For example, the external makers of the organs of sight and hearing, together with the organs of movement, begin the implementation of the predator's nutrition program (NP), realizing the finding and seizure of the victim – the object of the NP. After that, the external makers of the body's digestive system in the oral cavity – the jaws – carry out their fragmentation program, pre-grinding the object, preparing it to perform the main internal digestion program and subsequent assimilation of molecular food components by organ makers (OMs) of the stomach and intestines. After this, the transport blood supply program delivers these molecular components to each cell of the MO. All the makers of the mentioned programs that are part of a task-based food program make up the integrated maker of this task program.

Main and Auxiliary Task Programs

The division of programs into main (basic) and auxiliary seems justified. The main programs include those that solve local particular tasks in the course of executing the full task program, and, accordingly, we will consider as auxiliary programs (APs) that contribute to the implementation of basic ones or supplement them.

For example, the digestive program performed by the stomach and intestines is a part of the task program of nutrition and it is the main program, at the same time, the transport program implemented by the BSS is the AP.

Programs performed in the stomach by CMs, such as the parietal cells of the mucous coat that release the main components of the GJ which soften and decompose food fragments (FFs), are the main programs.

At the same time, the program to create a mucous-bicarbonate barrier that protects cells that are in contact with the lumen of the stomach from the action of GJ is logically attributed to auxiliary or additional.

Structure of MO Organs and Systems

The structure of organs and systems largely reflects the above principles for the structuring of complete and particular task programs and makers that execute them.

The systems that carry out internal programs include the digestive, respiratory, endocrine, sexual and immune systems which solve basic tasks, as well as the circulatory and NSs which perform auxiliary transport and information control programs.

For example, the digestive system contains makers of various levels; they carry out the basic programs of digestion and subsequent absorption of molecular food components into the blood. In addition, this system includes the site for the execution of these programs – the digestive tube. The essential elements of the digestive as well as all other systems and their constituent organs are peripheral sections of the transport circulatory and informational control NSs that could reach every cell.

It should be noted that if the basic systems are localized in the MO, then the auxiliary systems are delocalized and distributed throughout the organism.

The auxiliary system also includes the excretory system through which the waste products of the vital activity of the body are removed from the organism.

EPs are primarily performed by sensory and movement systems. The former includes makers – organs of sight, hearing, smell and touch that serve to collect and process information about objects of EPs and the situation in the virtual site of their execution in the external environment. The movement system includes organs for moving and positioning the organism, forgrabbing food and protection and it consist of muscular-articular makers of the limbs and jaws of the MO.

Movement programs are carried out by signals from the NS, after preliminary processing information about the state of the object and the environment coming from the sensory system.

EXAMPLES OF INTERNAL PROGRAMS

Let us consider in more detail some of the internal programs of a MO.

Nutrition Programs

The task program of cellular nutrition includes three local tasks and compound programs.

First, we start with the program of extraction in the environment of suitable food and its preliminary fragmentation, which is performed outside the MO, being thus external, and implemented by external organs – makers, which primarily include organs of movement, organs of capture and preliminary fragmentation of a food object, and sensory organs. We will look at this kind of EPs that determine the behavior of MO later in a separate section "External Programs".

The next component program is already internal, performed inside the MO. When executing this program, its makers, which make up the body's digestive system, carry out grinding and partial dissolution of FFs and, finally, bringing them to the level of the components consumed by the cells, proteins, amino acids, fatty acids, microelements, etc.

Last, the third final composite program, being a transport one, delivers the nutritional components consumed by cells from the digestive system into the extracellular space, from which the cells absorb them through their external endocytosis program. Makers of the MO's circulatory system perform this transport program.

In accordance with the definitions above, the first two programs are basic and the last transport program is auxiliary.

Let's consider in more detail the internal main program of transformation of a food object into the molecular components.

In the digestive system, conditionally, there are three main sections: anterior, middle and posterior.

The anterior section includes the organs of the oral cavity, the pharynx and the esophagus. In this section, mechanical food processing takes place.

The middle section consists of the stomach, small and large intestine, liver and pancreas. In this section, chemical processing of food, absorption of the products of its decomposition and the formation of feces happens.

The posterior part is represented by the caudal part of the rectum and provides the function of excreting undigested food debris from the alimentary canal.

At the same time, the oral cavity, pharynx, esophagus, stomach, small intestine and large intestine form successive sections of the digestive tube. The liver and pancreas are separately located organs.

The site for the implementation of the main programs included in the considered composite program for dissolving FFs is the inside of the digestive tube at the level of the stomach and small intestine.

The main actions that are performed with FFs in the stomach and small intestine are their dissolution and decomposition while stirring.

Programs for processing FFs in the stomach are carried out by GJ, the main components of which are hydrochloric acid and pepsin.

The acid has a bactericidal effect, it loosens the tissues, softens the fibers and promotes the conversion of the pepsinogen zymogen to the active enzyme – pepsin maker that breaks down proteins into shorter polypeptides and also turns new pepsinogen molecules into pepsin, implementing autocatalysis.

Special cells that make up the glands located in the mucosa produce and secrete GJ, which is the component of the complex maker, into the lumen of the stomach.

Thus, hydrochloric acid is formed from hydrogen and chlorine ions supplied to the gastric cavity by parietal cells.

The gastric chief cells synthesize and secrete the pepsin precursor – pepsinogen.

GJ is the result of social programs performed by a combination of glandular secretory cells that make up the epithelial tissue of the stomach. At the same time, GJ performs the main program for digestion of FFs outside of these cells in the lumen of the stomach, being a component maker of this program.

The chemical activity of GJ is so high that to protect against its action on the glands and other components of the stomach, the secret of specific mucous cells that form the so-called mucous-bicarbonate barrier, is used.

In addition to epithelial tissue, the composition of the gastric wall includes muscle formed by smooth muscle cells (SMCs). These tissues form layers with radial and longitudinal orientation, while they contract, they carry out programs for mixing the decomposable FFs from GJ into the so-called chyme and pushing this chyme along the lumen of the stomach and then into the duodenum – the beginning of the small intestine.

The muscle fibers that make up a muscle are made up of thin and thick filaments. These filaments make up a significant part of the volume of the muscle cell, which is an element of a tissue social maker.

The interacting proteins of thin and thick filaments were identified as actin and myosin, respectively. Actin, which is larger in the cytoskeleton than any other protein, forms structures capable of contraction together with myosin.

Each myosin molecule consists of six polypeptide chains – two identical heavy chains twisted into a so-called tail and two pairs of light chains forming the heads of the molecule.

The muscle shrinks as a result of the interaction of the myosin heads with actin filaments adjacent to them. During this interaction, the myosin heads hydrolyze adenosine triphosphate (ATP). The hydrolysis of ATP and the subsequent dissociation of the tightly bound hydrolysis products (ADP and P I) cause an ordered series of allosteric changes in the conformation of myosin. Some part of the energy being released turns into motor work. As a result, the myosin heads "walk" along the actin filament so that the thin filaments slip relatively thick one without changing the length of both of them and the muscle fiber is reduced.

Muscle contraction is triggered by an increase in the Ca^{2+} concentration in the cytosol of the muscle cell.

In order to interact with the actin filament, the myosin head must be phosphorylated.

This phosphorylation is catalyzed by a special enzyme which in turn becomes active only by binding to the complex of Ca^{2+} and calmodulin.

Thus, muscle contraction is controlled by the Ca^{2+} concentration in the cytosol of muscle cells.

In general, smooth muscle contraction is controlled by a large set of various signals, including impulses coming from the autonomic NS and hormones (e.g., adrenaline). Many of these signals act through the kinase of myosin light chains. In particular, adrenaline inhibits myosin light chain phosphorylation, causing SMCs to relax.

Homogeneous pre-processed food mixture – chyme is pushed in portions into the small intestine, where the components for consumption by cells are formed because of additional decomposition.

The wall of the small intestine is built from the mucous membrane, submucosa, muscular and serous membranes.

The inner surface of the small intestine has a characteristic relief of the villi and crypts. These structures increase the overall surface of the small intestine that contributes to the implementation of its main digestive functions.

The structural elements of all layers of the mucous membrane are involved in the formation of each villus.

Digestive processes occur in different zones of the intestine, and we can separate them to the processes of the extracellular digestion and intracellular digestion (in the cytoplasm of enterocytes). Extracellular digestion in turn includes cavitary (in the intestinal cavity), parietal (near the intestinal wall) and membrane (apical parts of the plasmalemma of enterocytes and their glycocalyx).

Enzymes of the digestive glands (salivary, pancreatic), enzymes of the intestinal flora and food enzymes carry out extracellular digestion in the intestinal cavity.

Parietal digestion occurs in the mucous sediments of the small intestine, which adsorb not only the various enzymes of abdominal digestion, but also the enzymes secreted by enterocytes. Membrane digestion occurs at the border of the extracellular and intracellular environment. On the plasmalemma and glycocalyx of the enterocytes, digestion is carried out by two groups of enzymes. The first group of enzymes is formed in the pancreas (a-amylase, lipase, trypsin, chymotrypsin, carboxypeptidase). They are adsorbed by glycocalyx and microvilli, with the majority of amylase and trypsin being adsorbed on the apical part of the microvilli, and chymotrypsin on the lateral zones. The second group – the enzymes of intestinal origin, they are associated with the plasmalemma of enterocytes.

Glycocalyx, in addition to adsorption of enzymes involved in digestion, plays the role of a filter that selectively transmits only those substances for which there are adequate enzymes. In addition, glycocalyx performs a protective function, ensuring the isolation of enterocytes from bacteria and the toxic substances formed by them. Glycocalyx contains receptors for hormones, antigens and toxins.

Intracellular digestion occurs inside the columnar epithelial cells, provided by their enzymes located in lysosomes.

Similar to the structure of the stomach, the intestinal wall is composed of muscle tissues formed by SMCs, ensuring the implementation of APs for mixing degradable FFs and pushing them along the intestinal lumen.

The monomers formed during the breakdown of proteins, carbohydrates and fats: amino acids, monosaccharides, monoglycerides and fatty acids – are then absorbed through the epithelial cells into the blood and lymph.

The Main Features of the Central Nutrition Program (CNP)

The program is carried out in the central medial organs of the digestive system – the stomach and small intestine, united in the digestive tube, with the participation of individual organs – the pancreas and the liver.

The result of the CNP are the monomers formed during the breakdown of proteins, carbohydrates and fats – amino acids, monosaccharides, monoglycerides and fatty acids absorbed into the blood and lymph.

Programs that make up the CNP are performed sequentially in the stomach and small intestine.

In the stomach, a compound maker of GJ carries out the main programs of softening and initial digestion of FFs in the lumen of this organ.

The main components of JS are hydrochloric acid (HCl), softening and grinding FF and inactive precursor of pepsin pepsinogen which degrades proteins by their hydrolysis. These are secreted into the lumen by two types of specialized cells: parietal and gastric chief cells, respectively. Activation of pepsinogen in the lumen of the stomach is triggered by HCl and it then proceeds autocatalytically: pepsin itself activates its preferment.

The AP to create a mucous-bicarbonate barrier protecting cells from the action of GJ is performed by specific mucous cells secreting mucin, mucopolysaccharides, gastro-mucoprotein and bicarbonate, the mixture of which forms a collective protective marker.

APs performed by SMCs include mixing FF and pushing them to the digestive organ next to the stomach, the small intestine. The social makers of these cells that perform the necessary reductions for the mechanical action are muscle fibers consisting of interacting protein molecules of actin and myosin.

In the gastric programs discussed above, two types of social makers can be distinguished. Some of them, being secreted from the corresponding specialized cells, carry out programs outside these cells, and we will call them external. Others perform mechanical programs, remaining inside the muscle cells and thus are internal.

In both cases, specialized cells generate inherent social makers, spending their material and energy resources.

The life span of gastric secretory cells is about 3 days. Their regeneration occurs at the expense of stem cells located at the bottom of the gastric pits, by mitotic division of progenitor cells.

The replenishment of the material and energy resources of cells – micromolecules and ions, including oxygen for respiration – is carried out from the capillaries through the intercellular space.

The structure of the stomach – the structure of the location of the tissue corresponds to the programs they perform.

Site to perform the main program – the lumen of the gastric cavity.

The secretory tissue leaving the lumen forms the inner layer of the gastric wall – the mucous membrane.

This layer is formed by secretory parietal, gastric chief and mucous cells, through the apical ends of which the makers, which constitute the GJ and form the mucus-bicarbonate barrier, enter the lumen of the stomach.

The opposite basal end of these cells enters the next layer of the wall – the submucosa. Blood, lymphatic vessels and nerve endings penetrate this layer consisting of loose connective tissue. Oxygen, nutrients and humoral factors that control intracellular makers are delivered from the circulatory system and through tissue fluid, which is also located in the submucosa. Also, neurotransmitters enter the cells through tissue fluid, transmitting cell control signals from the NS.

The third muscular layer of the stomach wall consisting of two sublayers is adjacent to the submucosa. One of them is internally composed of SMCs, oriented radially and providing mixing of soluble FFs, while the second external one contains longitudinally oriented SMCs that move the chyme along the gastric lumen and further into the small intestine. Nutrition and innervation of muscle tissue realize through tissue fluid submucosa.

Finally, the fourth outer serous membrane faces the abdominal cavity. It has good transport properties, for example, for suction of excess liquid from the cavity and ensures unhindered sliding of the organs in the cavity relative to each other.

Thus, the structure of the stomach provides the effective performance of the main and APs of this organ by specialized cells and their makers. Additionally, these cells are supplied with oxygen, essential nutrients, signal ions and molecules.

The same principles used in the structure of other organs of the digestive system, digestive tube with organs of the oral cavity, pharynx and esophagus, which make up its anterior section, as well as intestines forming its middle and posterior sections and located outside its glands: salivary glands, liver and pancreas.

In addition, each body has a specificity determined by local programs, the implementation of which it provides.

In the small intestine, the organ following in the digestive tube behind the stomach, the central compound nutritional program continues to be implemented, which splits food into the level of fragments consumed by cells (CC) and their subsequent absorption into the blood.

To the main program of further decomposition of FFs here is added a program for transferring CC fragments to tissue fluid, through which they enter the blood.

Epithelial cells, also known as enterocytes, perform the programs of the decomposition of FFs to the CC state and their subsequent transportation into the tissue fluid of the submucosa. In the cytosol of the enterocytes and on the apical sides of their membranes, the programs are performed by internal EMs of these specialized cells. In the parietal space and in the intestinal cavity, the programs are carried out by external EM secreted by enterocytes. External EMs, secreted by specialized cells of the pancreas and liver and entering the intestine through the ducts of these glandular organs, are added to them. All of these external makers form intestinal juice that digests the chyme that comes from the stomach, right up to CC molecular fragments of all types of nutrients: proteins, fats and carbohydrates. While in the process of digestion, all these CC components are absorbed by intestinal epithelial cells and then enter first into the tissue fluid and after that, they get into the blood.

Like the stomach, the small intestine has four membranes: mucous, submucous, muscular and serous with the same basic functions.

Additional programs of the small intestine include the following.

First, they include the programs of local control of the intestine's main programs, performed by external makers – hormones and biologically active amines secreted by endocrinocytes.

Further, it is a program of moistening the surface of the intestinal mucosa and thereby contributing to the mixing and advancement of food particles, performed by mucus, secreted by exocrinocytes.

And, finally, the programs performed by specialized cells of the muscular layer consist in mixing and pushing the chyme along the intestine.

Monomers – amino acids, monosaccharides, monoglycerides and fatty acids – formed during the breakdown of proteins, carbohydrates and fats are then absorbed through the epithelial cells into the blood and lymph, where they are the objects of auxiliary transport program that delivers these components to each cell through the intercellular space.

Breathing Programs

The main task program of respiration is to provide oxygen for the cellular energy programs that creates high-energy molecules and ionic gradients through the process of the nutrient's oxidation. Two complex system makers (SMs), body's respiratory and circulatory systems perform this task program.

The same organismic systems perform an auxiliary task respiratory program that removes the end product of intracellular energy programs from the body – carbon dioxide.

The integral program of supplying all cells of the body with oxygen includes four-phase programs. During the first phase, the oxygen in the composition of the ambient air during inhalation is carried by convection transport into the ventilation part of the respiratory system – pulmonary alveoli. In the second phase, oxygen diffusively enters from the alveoli into the nearby red blood cells located in the capillaries of the circulatory system. In the third phase, oxygen in the erythrocytes is convectively transferred by blood to the capillaries of the tissues. Finally, the oxygen supply program ends with diffusion of oxygen from the capillaries into the surrounding cells of the body tissues.

The carbon dioxide emitted as a result of the implementation of energy programs by cells is removed from the body by similar phase programs implemented in the reverse order. During the first phase of the program, the carbon dioxide of the tissue cells diffuses into the closely located capillaries, where it is bound by red blood cells. In the second phase, bound carbon dioxide from the blood stream is transferred to the alveolar capillaries. In the third phase, the liberated carbon dioxide diffuses into the alveoli, after which during the fourth phase it is expelled by the respiratory tract from the body.

Let us consider these phase programs in more detail, we will begin with the local task program of supplying cells with oxygen.

The first phase of air drift into the ventilating alveolar part of the respiratory system is performed through the organs of this system, subdivided into upper and lower respiratory tract. The upper respiratory tract includes the cavities of the nose, nasopharynx and oropharynx, the lower – the larynx, the trachea and the bronchi, passing into the lungs containing alveoli.

Air enters the body through two nostrils, each of which is lined with epithelium, in which there are cells secreting mucus that trap dust particles. Also, the mucus moistens the inhaled air and in the nasal passages, the air heats up due to shallow vessels. After passing through the nasal passages, air enters the throat through two internal holes. The slit-like opening leading to the larynx is protected from food by the triangular valve of cartilage tissue – the epiglottis.

The larynx is a cavity in front of the entrance to the trachea, formed by nine cartilages. The muscles attached to them allow these cartilages to move relative to each other.

From the larynx, air enters the trachea – a tube that lies directly in front of the esophagus and ends in the cavity. The walls of the trachea are reinforced with cartilages that do not allow the trachea to fall down when inhaling. Inside the trachea is lined with epithelium that contains secreting mucus cells. Moisture particles and microbes that are trapped in the trachea get in the mucus, and the rhythmic beats of the cilia directed toward the oral cavity remove them from the trachea.

At the lower end, the trachea is divided into two bronchi. The right bronchus of the trachea is divided into three smaller bronchus, each of which is directed into one of the lobes of the right lung. The left bronchus of the trachea is divided into two bronchi, which end in two lobes of the left lung. In both lungs, each bronchus is divided several times into even thinner tubes, called bronchioles, in which the wall that consists of smooth muscles, connective tissue with elastic fibers, providing the possibility of stretching and narrowing of the bronchioles, and their ciliary epithelium with secreting mucus cells. The smallest tubes, called bronchioles, are about 0.5 mm in diameter. They, in turn, are divided into numerous alveolar passages lined with cubic epithelium and ending in alveolar sacs, called alveoli. The alveoli create the surface on which gas exchange occurs.

The movement of air inside the alveolar pulmonary part of the respiratory system during inhalation and in the opposite direction from the lungs to the outside during expiration is carried out by convection due to air pressure drops in the lungs relative to external pressure. When you inhale, the drop is provided by an increase in the volume of the lungs, and when you exhale, it decreases. In turn, the change in lung volume occurs due to changes in the size of the chest, carried out with the help of respiratory muscles.

The change in the volume of the chest is carried out with the raising and lowering of the ribs with simultaneous contraction of the diaphragm.

In the covered above program phase of respiration, air drift into the pulmonary ventilation part of the respiratory system, the main makers are precisely the ribs and diaphragm together with the corresponding muscles adjacent to them, which create pressure drops in this system necessary for convection air movement.

The airways limit and direct the movement of air molecules at the inhalation and exhalation.

At the same time, for example, the epithelial tissues of the respiratory tract are a collective cellular maker that performs an important AP for cleaning the incoming air of solid particles and microbes by excreting mucus.

In the second phase, oxygen diffusively passes from the alveoli into the CM erythrocytes in their vicinity, located in the capillaries of the circulatory system and performing a transport program for the transfer of oxygen in the third phase to the tissue cells.

In order for the exchange by diffusion to be sufficiently effective, the exchange surface must be large and the diffusion distance small. The diffusion barrier in the lungs fully meets these conditions. The total surface area of the alveoli is about 50–80 m^2. According to its structural characteristics, the lung tissue is suitable for diffusion: the blood of the pulmonary capillaries is separated from the alveolar space by a thinnest layer of tissue. In the process of diffusion, oxygen passes through the alveolar epithelium, the interstitial space between the main membranes, the capillary endothelium, the blood plasma, the erythrocyte membrane and at last the erythrocyte internal environment. The total diffuse distance is only about 1 micron.

The outer side of the alveolar wall is covered with a dense network of blood capillaries, the diameter of which is less than the diameter of red blood cells, and red blood cells squeeze through them under the pressure of the blood. At the same time, a large proportion of their surface comes into contact with the surface of the alveoli, where gas exchange takes place, and more oxygen enters the red blood cells. In addition, erythrocytes move relatively slowly through the capillary, so that gas exchange may take longer.

In highly specialized CMs of erythrocytes, oxygen is captured by an elementary protein maker by hemoglobin (Hb), which accounts for ≈98% of the mass of erythrocyte cytoplasm proteins. A feature of oxygen binding to hemoglobin is its allosteric regulation – the stability of oxyhemoglobin falls in the presence of 2,3-diphosphoglyceric acid, an intermediate product of glycolysis and, to a lesser extent, carbon dioxide, which contributes to the release of oxygen in the tissues that need it.

An important role in the erythrocyte is played by the cellular (plasma) membrane that transmits gases (oxygen, carbon dioxide), ions (Na, K) and water.

The erythrocyte formation program (erythropoiesis), auxiliary to the respiratory program, occurs in the bone marrow of the skull, ribs and spine. The life span of an erythrocyte is 3–4 months. Programs of their destruction (hemolysis) are implemented in the liver and spleen. Before entering the blood, the red blood cells pass through several stages of proliferation and differentiation in the composition of the erythron – the red hemopoietic germ.

During the third phase, the compound maker of the BSS performs the transport program for the transport of oxygen in the composition of oxyhemoglobin inside the erythrocytes.

In the fourth phase, when erythrocytes containing oxyhemoglobin are released into tissues depleted in oxygen and saturated with carbon dioxide, oxygen is released from the elementary oxyhemoglobin maker as a result of the above allosteric modification of this molecule. The released oxygen diffuses through the cell membrane of the erythrocyte, the extracellular space and the plasma membrane of the target cell (consumer) and enters this cell, where it is captured by cytochromes – the elementary intracellular makers that perform the cell respiration program, which we considered in Chapter 1.

The task program for removal of carbon dioxide from the body includes similar phases, but they are performed in the reverse order.

At first, the carbon dioxide emitted by the cells is diffusedly transferred into the capillaries in the immediate vicinity of these cells. Here, a certain amount of carbon dioxide remains in a state of physical dissolution, but most of it undergoes a series of chemical transformations.

First, the carbon dioxide molecule hydrates to form an acid. In the blood plasma, this reaction proceeds very slowly but in the erythrocyte it accelerates about 10 thousand times. This is due to the action of the enzyme carbonic anhydrase, which is an internal elemental maker of the erythrocyte. The next reaction is the dissociation of weak carbonic acid into bicarbonate ions and hydrogen ions.

Most of the carbon dioxide is transported in the form of bicarbonates and carbamino compounds, in the blood plasma and inside the erythrocytes during the implementation of the circulatory system program.

Molecular carbon dioxide, which is physically dissolved in the blood plasma, diffuses from the blood into the alveoli. In addition, carbon dioxide which is released from the carbamino compounds of erythrocytes due to the oxidation of hemoglobin in lung capillaries, as well as from blood plasma bicarbonates as a result of their rapid dissociation by the enzyme carbonic anhydrase contained in erythrocytes also diffuses into the alveoli of the lungs.

Finally, the last phase of carbon dioxide removal is carried out by the convection method through the respiratory tract at the exhalation stage, when the volume of the chest and consequently the lungs decreases, and the pressure in them increases.

The main programs that control breathing perform a rhythmic alternation of inhalation and exhalation. Several groups of nerve CMs are involved in the execution of these programs. For the consistent implementation of inhalation and exhalation is responsible "center", located in the lower sections of the brain stem. At the same time, an isolated medulla is able to generate a respiratory rhythm. However, the stabilization and coordination of this rhythm requires the participation of nerve formations located in the middle and upper sections of the brainstem bridge. It was shown that in two limited areas of the medulla oblongata, neurons are located on both sides, discharging shortly before inhalation and during the inhalation itself.

In the area located along the nucleus, caudal to the region, were detected neurons that are discharged during both exhalation and respiratory pause.

In the medial inspiratory region, both neurons excited by inhalation and neurons that are discharged not only simultaneously with alpha-neurons, but also during their pause were detected along the single tract. The activity of beta-neurons increases with maximum stretching of the lungs, so they are thought to have an inhibitory effect on alpha-neurons.

Thus, the rhythmic alternation of inhalation and exhalation is associated with alternating discharges of inspiratory and expiratory neurons. During the activity of inspiratory neurons, expiratory neurons are not discharged, and vice versa. This suggests that inspiratory and expiratory neurons have a reciprocal inhibitory effect on each other.

Inspiratory and expiratory neurons perform informational control programs for the innervation of the respiratory muscles that perform motor programs to change the volume of the chest, which results in change of the lungs' volume.

Immune Programs

Tasks and Structure

The most important group of MO programs performs the basic task of protecting against microorganisms – immune protection. In the course of its solution, information programs are carried out to detect, identify microorganisms and then effector programs to destroy them and/or body cells affected by them.

IPs form an extensive network of local task programs performed by makers of different levels of the hierarchy: EMs – proteins, CMs and OMs.

Considering the complex system of structural and organizational sense of the IP, we will try to track, on its example, some work principles and interaction of makers of different levels in the implementation of their overall task program.

We recall that the selection of makers of a more complex structural hierarchy into a separate group is due to the fact that they carry out specific constituent programs using their constituent EM (proteins, RNA and complexes of them), interacting with objects and other makers as a whole.

At the same time, the structural makers of the CM and the OM perform important functions for the makers of the lower hierarchy: the CM for the EM and the OM for the CM. They include sites for performing basic programs of junior, lower makers, and also contain auxiliary makers that help reproduction, proliferation and/or maturation, formation of main makers.

There are several options, modifications of IP depending on the following factors: the primary or secondary nature of the appearance of PMs in the MO, the scale degree of infection of the MO by microorganisms as well as the localization of PM site of in the MO.

The main division into two directions of IP innate and adaptive immune responses is carried out according to the methods of recognition of PMs and molecules.

Two types of the CMs perform pathogen recognition information programs: myeloid cells and lymphocytes. Both of these cell types are leukocytes.

Myeloid cells, which are primarily macrophages and dendritic cells (DCs), recognize with their receptor EM the so-called pathogenicity patterns (pathogen-associated molecular patterns – PAMP) – groups of molecules characteristic of all major pathogens (viruses, bacteria, fungi, protozoa, parasites). This happens at the early first stage of MO infection, when the PAMP receptors, which respond to all possible types of pathogens, produce signals that trigger the innate immune response. Thus, the receptor EM myeloid CM give a signal about the presence of an invasion into the MO of any PM, without isolating (determining) its specific species.

On the other hand, many types of receptors B- and T-lymphocyte CM are able to recognize specific pathogen by its specific molecule – the antigen.

An important feature of these receptors is the gigantic variability of their antigen-recognition domains (millions of variants within one organism).

All variants of antigen-recognition receptors are distributed between cells clonal.

Recognition of antigens occurs individually (and not in groups, as in the case of PAMP) and starts the second direction of IP – an adaptive immune response.

Adaptive immunity (AI) programs work with specific pathogens that attack the MO. This allows for more targeted and efficient use of the body's resources.

Implementation Examples

At the beginning of the PM attack starts a program of innate immunity (II).

The main cellular makers of II are myeloid macrophage cells (MCs). Macrophages exist in two varieties: resident and inflammatory cells. Resident macrophages that colonize tissues, primarily barrier ones, are the first to recognize the PM by its PAMP, carrying out its main information program. In turn, inflammatory MC represents an additional set of macrophages, mobilized during the implementation of the immune response to combat pathogenic attack.

The signals produced by MC's receptors activate the MC and they begin to carry out producing and secreting programs into the intercellular space with a further entry into the blood and lymphatic systems of cytokines – elementary informational protein makers that play a key role in IPs.

Different types of cytokines, acting on the receptors of the respective CMs, launch in the region of penetration of pathogens into the organism, the so-called focus of inflammation, important auxiliary immune response programs as proliferation programs and subsequent maturation of CMs, first inflammatory MCs, and transport programs of these CMs.

The receptors of the MC that have fallen into the inflammation center are likely to detect the pathogen there and start the effector program phagocytosis, which results in macrophage's absorption and destruction of the pathogen microorganism.

Let us consider some features of auxiliary and additional programs that ensure that macrophages efficiently carry out their basic immune response programs: information ones – the detection of a PM and the effector – its destruction.

The most significant APs are reproduction programs in the required quantities of MC makers and their transport to the inflammatory focus, i.e., pathogen attack site.

The initial object for the program of production of CMs – macrophages – are hematopoietic stem cells localized in the bone marrow, which is the main OM of the immunity task program. These cells implement the division program approximately once every 30–60 days. Stem cells enter the cell cycle and then begin to differentiate when contacts with osteoblasts are lost.

At the same time, there is a balance of programs of proliferation and differentiation: at the population level, one of the daughter cells continues to divide, while the other undergoes differentiation, that is, it matures.

To differentiate myeloid (as well as any other) cells, it is necessary to launch programs that reproduce the corresponding social makers. In turn, this implies the expression of certain genes, which is provided by the program of activation of the promoter regions of these genes. In myelopoiesis, this requires, in particular, a high level of expression of the transcription factor PU.1, participating as one of the makers in such an activation program.

Immature hematopoietic cells easily go through the process of apoptosis. To maintain their viability, they need the presence of colony-stimulating factors (CSFs) in the microenvironment, which are cytokines. Granulocyte-macrophage CSF (GM-CSF) is considered the main cytokine common to almost all myeloid cells, starting from a common myeloid precursor.

Thus, primary production and secretion of cytokines by macrophages, in response to recognition of the pathogen, leads, in particular, to an increase in the number of these CMs by reducing the effectiveness of apoptosis of immature cells.

After maturation, the monocytes stay in the bone marrow for another day and then leave it, entering the bloodstream. At the same time, cells retain the ability to divide and further differentiate.

Monocytes and macrophages represent the stages of development of myeloid cells. They form a mononuclear phagocytic system.

The circulating cell variant is a monocyte, while the tissue variant is a macrophage. The transformation of a monocyte into a macrophage occurs under the influence of the tissue microenvironment and is accompanied by the expression of new genes, i.e., can be considered as a continuation of the cell differentiation program. This differentiation is regulated by the colony-stimulating factor M-CSF.

Transport programs that develop in stages create the path of monocytes from the bone marrow to the colonization tissue, including in the inflammatory focus.

At the first stage, a program is implemented to release leukocytes from the bone marrow into the bloodstream.

The release of leukocytes from the bone marrow into the bloodstream occurs due to the weakening of the interaction of chemokines (a type of cytokine) secreted by the stromal cells of the bone marrow with the receptors of leukocytes. The most important chemokine that retains maturing cells in the bone marrow is CXC12, recognized by the receptor for leukocyte CXCR4. Under the influence of CSFs (hematopoietin), the production of chemokines and the expression of their receptors are weakened, which allows mature cells to leave the bone marrow and then get into the bloodstream.

The next important stage of the transport program is the emigration of leukocytes from the bloodstream. Movement in the channel by rolling the cells along the capillary or venule walls in the direction of the blood flow stops after receiving signals for the endothelial cells of the vessel walls and leukocytes from cytokines secreted by macrophages. After the actions of these signals, there is a cardinal enhancement of the interaction between leukocytes and these endothelial cells and, as a result, the fixation of the leukocyte on the vascular wall.

The subsequent advance of the leukocyte between endothelial cells is carried out with the help of PeCAM makers – in the penetration of the lymphocyte into an endothelium and exit from it and CD99 – in overcoming the contact zone between the endothelial cells.

A chemotaxis program carries out the last stage of the task of the transport program – direct advance to the zone of inflammation.

In our case, the term chemotaxis is used to refer to the movement of a leukocyte cell in the direction of the gradient of chemotactic agents emanating from the focus of inflammation. The main types of such agents (recognized by receptors of macrophages or their monocyte precursors) are bacterial peptides fMLP and chemokines of the methyl-accepting chemotaxis protein (MCP) and MIP groups.

The condition for directed cell movement is the presence of a concentration difference of chemotactic agents of at least 1% over a distance equal to the diameter of the cell.

At the basis of the movement of leukocytes lays the reaction of the cytoskeleton's contractile proteins, primarily actin.

Mobilization of cytoskeletal components occurs when chemotactic factors are activated via rhodopsin-like receptors. Due to the polarization and the reorganization of the cytoskeleton that occurs during this process, the round-shaped cell becomes triangular. In the direction of the object of chemotaxis, lamelopodium is pushed – a portion of the cytoplasm that is poor in organelles, but contains a network of microfilaments, in particular, filamentous F-actin.

Cell orientation in the process of chemotaxis is determined by the polymerization of microtubules. The process of movement is the reduction of microfilaments.

A macrophage formed from a monocyte that has got into the focus of inflammation, i.e., inflammatory macrophage begins the implementation of its main effector program phagocytosis for the destruction of the PM.

Due to the preceding chemotaxis, by the beginning of phagocytosis, the macrophage turns out to be polarized: the cytoskeleton threads and the organelles are oriented toward the source of chemotactic signals – the object of phagocytosis, and the membrane molecules necessary for phagocytosis are located on the pole of the cell facing the target.

The program of phagocytosis is carried out in several usually distinguished stages.

First, a macrophage approaches a specific pathogenic object when performing a chemotaxis program with orientation, for example, according to the fMLP bacterial chemotactic agent.

Next, an adhesion stage is carried out. The macrophage interacts physically with the PM, in which adhesive EMs participate, in particular, β1-integrin VLA-4 and β2-integrin LFA-1.

At the initial stages of phagocytosis, the main events take place on the polarized macrophage portion facing the PM where a temporary structure, called the phagocytic cup, should form. Polymerization of actin in combination with immersion of PM inside the leukocyte is the basis for the formation of this bowl – the starting position of the next stage of phagocytosis.

The immersion of PM is due to the reduction of the actin threads, concentrated around the phagocytic bowl. Immersion of the forming phagosome into the cell is completed by closing the membrane over it, like a zipper.

Moving inside the cell, the phagosome undergoes a process of maturation. It is based on multiple acts of fusion with the phagosome of various granules that introduce effector molecules into it.

Lysosomes make a decisive contribution to phagosome maturation and their ability to kill and split absorbed objects. The merger of phagosomes and lysosomes is considered as the beginning of the phagolysosome formation.

The local transport program for the delivery of NADPH oxidase complexes (which is performed with the participation of specific granules) is the most important program for the maturing phagolysosome. As the contents of various granules are sequentially infused into the phagolysosome, the pH of its contents changes, its bactericidal potential increases, as well as its ability to destroy certain substrates, in this case PM.

The formed phagolysosome or cellular organelle is a compound maker that performs the final effector program of phagocytosis – killing and cleavage of PM.

The effectiveness of phagocytosis is caused not so much by the absorption of the pathogen as by its destruction within the cell.

Killing takes place in phagocytic phagolysosomes that contain chemical agents formed by the intracellular macrophage makers. An example of such agents that destroy microorganisms is hydrogen peroxide, and especially the hydroxyl radical, which has a very strong bactericidal activity. With their joint action occurs lipid peroxidation, disruption of peptide bonds, oxidation of sulfhydryl groups and other deep chemical changes of macromolecules in the cell walls of pathogens, all those processes lead to their death.

The considered branch of the II program with macrophages as the main CMs is not the only one. Neutrophils, a type of myeloid cell and protein elemental maker (EM), components of the so-called complement also carry out their II programs effectively.

The main makers of AI programs, for example, T-lymphocyte cells, also take part in II programs.

AI programs are usually launched at a later stage of the pathogenic attack, when a failure of II is detected.

These programs begin with the identification and presentation of the individual information characteristic – the antigen of a particular attacking microorganism for T- and B-lymphocytes – the main cellular makers of the AI.

This initial program is being performed by the DCs that are the kind of myeloid cells.

At the same time interacting in barrier tissues with pathogens, DCs, through their receptors, carry out an information program recognizing PAMP of PM and then absorb it using various endocytosis programs.

The considered examples of IPs are consciously limited by the author both in terms of the breadth of coverage of various options for their implementation and in details of their realization and serve the purpose of formulating the main features of such programs.

Some Characteristics

As already noted at the beginning of this section, when solving a common task of IP, fighting PMs attacking a MO, two local, specific task programs are executed: informational – pathogen detection and effector one – on its destruction and/or on reducing its effectiveness in MO.

Both of these programs are major and are performed by the base makers. In the considered examples of II, the role of such makers is performed by cellular makers (CM), macrophages and EMs cytokines. In the case of AI, these are cellular makers: DCs, T- and B-lymphocytes, as well as EMs: specific antigen receptors of T- and B-lymphocytes, antibodies (they are immunoglobulins) and also, components of complement, cytokines.

As usual, for the solution of problems, general and particular, the role of APs is great.

In the pool of IPs, the result of the AP implementation is primarily the mobilization of the required number of basic makers and their delivery to the sites of their respective programs.

For this, it is necessary to carry out a number of concrete composite programs of the first level, both effector and informational.

The effector APs include programs for the proliferation, differentiation and transport of cellular base makers II and AI.

The second is programs for generating and recording signals that trigger the next stages of effector programs. Typical examples of such auxiliary information programs are the production and secretion by macrophages of information EMs cytokine and the subsequent program of interaction of these EMs with CM receptors, after which the activation signal is generated, i.e., launch of appropriate effector programs.

In programs of II, the MC is a universal CM that performs both basic and APs related to both effector and informational varieties. Thus, a resident MC, located in MO tissues attacked by a pathogen, recognizes the pathogen through its EM receptors by performing basic informational programs. After that, the MC is activated and performs an auxiliary information program for the production and secretion of elementary cytokine makers. EM cytokines are the main "assistants" of macrophages, performing activation programs of CMs, participants of II programs.

A feature of the IPs, both basic and auxiliary, is in most cases the absence of specific dedicated sites. These programs are performed in all tissues and parts of the humoral systems of the body that are attacked by pathogens. In this regard, in the absence of a pronounced attack in different parts of the MO, there are duty cell and elementary immune makers. Those of them that are localized in the tissues are called resident. The role of these makers is to detect the beginning of an attack and react to it by launching primarily programs for mobilizing and activating the composition and number of main and auxiliary makers that are necessary for a specific attack, considering its localization, intensity and type of pathogens participating in it.

At the same time, a specific area that pathogens attack in MO becomes a virtual site for the execution of IPs, the so-called inflammation zone, the distinguishing features of which are, in particular, the enhancement of the local blood circulation and gradient trajectories marked by chemokines to attract mobilized inflammatory cellular immune makers to this zone.

However, some parts of the IPs are performed in specific sites of the hematopoietic and lymphatic organs. So, the initial programs for producing CMs – macrophages from hematopoietic stem cells – are performed in the bone marrow, which can be considered as one of the immune OMs. In the T-zone of the lymphoid organs, programs of final differentiation (maturation) of DCs are implemented, and then information programs on presenting the antigen to T-lymphocytes.

The IP pool also has features for managing and synchronizing its constituent programs. Much of the control programs are run by information programs within this pool. And as already noted, cytokines plays the crucial role in there.

The immune system, despite being autonomous, is under the control of endocrine and nervous effects. The effect of hormones and mediators of the autonomic NS is realized through their interaction with specific cells receptors of the immune system.

The immune response requires reliable control and regulation, which, on the one hand, should ensure its effectiveness, and on the other, limit it for the safety of the body.

Transport Programs

Among the tasks solved by the transport programs of a MO and which are cardinal for its social structure, we will examine several of the most significant.

First, this is the distribution of nutrients and oxygen necessary for cellular nutrition and respiration, in other words, objects of intracellular programs.

Next, it is the transport of elementary and cellular makers from the sites of their creation to the sites of their further differentiation or the execution of corresponding programs by them.

In addition, this is the realization of the so-called humoral factors of controlling many MO programs of the endocrine primarily of the hormonal system of the body by delivering hormones, biologically active substances, nonorganic metabolites and ions to the controlled CMs.

Two transport systems help solving those problems – the BSS and the lymphatic system.

Blood Supply System – Basic Transport System of MOS

In the blood supply system (BSS), the listed objects are transported by blood – a liquid substance consisting of water and these objects in suspension. The blood moves along a closed circuit of vessels, hollow tubes that penetrate all organs and tissues of the body due to periodic contractions of the central organ of the BSS – the heart.

BSS solves two local problems. The first is to provide convection movement in the directions to the cells and from the cells of the MO of the above-mentioned makers and program objects. The second is the exchange or facilitation of the exchange of these makers and objects between the blood and the intercellular interstitial fluid, the external exchange medium for the body's cells.

Heart

The heart is the main maker performing the program of convective movement of blood through the vascular system – the site of the transport program.

The heart is a hollow muscular organ that pumps blood through the vessels with a series of contractions and relaxations. Depending on the species, the heart inside can be divided by partitions into two, three or four chambers. In mammals and birds, the heart is four chambered. The chambers are distinguished on a blood flow as right auricle, right ventricle, left auricle and left ventricle. Auricles and ventricles do not interconnect.

The wall of the heart has three layers: the inner endocardium, the outgrowths of which form valves; the middle one is the myocardium, the heart muscle, the contraction of which causes the execution of the pumping program; and outer epicardium covers the surface of the heart.

The pumping function of the heart is based on the alternation of relaxation (diastole) and contraction (systole) of the ventricles of the heart. During diastole, the ventricles fill with blood, and during systole, they throw it into the arteries of the aorta and pulmonary trunk. At the exit from the ventricles are located valves that prevent the flow of blood from the arteries into the heart. Before filling the ventricles, blood flows through the large veins (hollow veins and pulmonary veins) into the atria. Atrial systole precedes ventricular systole, thus, the atria are, as it were, auxiliary pumps contributing to the filling of the ventricles.

The vessels that carry blood from the heart to the organs are called arteries, and from organs to the heart they are called veins.

Among the vessels of the circulatory system are arteries, arterioles, precapillaries, capillaries, postcapillaries, venules, veins.

There is no gas exchange and diffusion of nutrients in the arteries and veins, there is only a delivery route. As the blood vessels move away from the heart, they become smaller.

Between the arteries and veins is a microcirculatory bed that forms the peripheral part of the cardiovascular system. The microvasculature is a system of small vessels, including arterioles, capillaries, venules, as well as arterio-venular anastomoses. Here occur metabolic processes between the blood and tissues.

Vessel Structure and Types

The wall of a blood vessel consists of several layers: the inner (tunica intima), containing endothelium, sub-endothelial layer and the inner elastic membrane; medium (tunica media), formed by SMCs and elastic fibers; outer (tunica externa), represented by loose connective tissue that contains the nerve plexus and vasa vasorum. The wall of a blood vessel receives nourishment at the expense of branches extending from the main trunk of the same or the adjacent artery. These branches penetrate the wall of the artery or vein through the outer sheath, forming a plexus of arteries in it, so they are called "vessels of blood vessels" (vasa vasorum).

Arteries gradually transform into arterioles, from which the microcirculatory vascular bed (MVB) begins. The MVB includes arterioles, precapillaries (precapillary arterioles), hemocapillaries, postcapillaries (postcapillary venules) and venules.

Arterioles are small vessels with a diameter of 50–100 microns, gradually turning into the capillaries. The main function of arterioles is to regulate blood flow to the main metabolic link of the MVB – hemocapillaries. In their wall all three shells, characteristic of larger vessels, remain, although they become very thin.

The internal lumen of the arterioles is lined with endothelium, under which located single cells of the sub-endothelial layer and a thin internal elastic membrane.

Smooth myocytes (muscle cells) are spirally arranged in the middle shell. They form only one to two layers. SMCs have direct contact with endotheliocytes, due to the presence of perforations in the inner elastic membrane and in the basal membrane of the endothelium. Endothelium-myocyte contacts provide signaling from endotheliocytes that perceive changes in the concentrations of biologically active compounds that regulate the tone of arterioles, to SMCs. Arterioles have a pronounced contractile activity called vasomotion. The outer sheath of arterioles is extremely thin and merges with the surrounding connective tissue.

Precapillaries (precapillary arterioles) are thin microvessels (about 15 microns in diameter) extending from arterioles and passing into hemocapillaries. Their wall consists of the endothelium lying on the basement membrane, SMCs located alone and the outer adventitial cells. In the places of discharge from the capillary arterioles of the blood capillaries, there are smooth muscle valve sphincters that regulate blood flow to certain groups of hemocapillaries and in the absence of a pronounced functional load on the organ most of the capillary sphincters are closed. In the sphincter region, smooth myocytes form several circular layers. Endotheliocytes have a large number of chemoreceptors and form many contacts with myocytes. These features of the structure allow precapillary sphincters to respond to the action of biologically active compounds and change blood flow to the hemocapillaries.

Hemocapillaries are thinnest walled vessels of the microcirculatory bed, through which blood is transported from the arterial to the venous. The functional importance of hemocapillaries is extremely high. They provide directional movement of blood and metabolic processes between the blood and tissues. According to the diameter, hemocapillaries are subdivided into narrow (5–7 microns), wide (8–12 microns), sinusoidal (20–30 microns and more with a diameter changing along the way) and lacunae.

The wall of the blood capillaries consists mainly of cells – endotheliocytes, as well as a non-cellular component – the basement membrane. Outside, a network of reticular fibers surrounds the capillaries. A single layer of flat endothelial cells forms the inner lining of hemocapillaries. The wall of the capillary in diameter forms one to four cells. Endotheliocytes have a polygonal shape, containing, as a rule, one nucleus and all organelles.

The most characteristic ultrastructure of their cytoplasm is pinocytotic vesicles. The latter are especially numerous in thin peripheral (marginal) parts of cells. Pinocytic vesicles are associated with the outer (luminal) and inner (abluminal) surfaces of the endotheliocyte plasmalemma. Their formation reflects the process of trans-endothelial substance transfer. At the confluence of pinocytotic vesicles, continuous trans-endothelial tubules are formed.

Plasmalemma of the luminal surface of endothelial cells is coated with glycocalyx that performs the function of active absorption of metabolic products and metabolites from the blood. In the endothelium of the hemocapillaries, there are "holes" (fenestrae) with a diameter of about 50–65 nm, closed by a diaphragm 4–6 nm thick. Their presence facilitates the course of metabolic processes.

Endothelial cells possess dynamic adhesion and continuously slide one relative to another. Between the endotheliocytes in the hemocapillaries of some organs, slit-shaped pores and an intermittent basement membrane are found. These intercellular spaces are another way of transporting substances between the blood and tissues.

There are three types of capillaries: (1) continuous, or somatic capillaries, located in the brain, muscles, skin; (2) fenestrated, or capillaries of the visceral type, located in the endocrine organs, kidneys, gastrointestinal tract; (3) intermittent, or capillaries of sinusoidal type, located in the spleen, liver.

In the hemocapillaries of the somatic type, endotheliocytes are connected to each other by means of tight contacts and form a continuous lining. The presence of such capillaries with a solid endothelial lining in the brain is necessary for the reliability of the blood-brain barrier.

The number of hemocapillaries in different organs varies. For example, there are up to 400 capillaries per 1 mm^2 of area in a cross-section in a muscle, whereas in the skin there are only 40. Under normal physiological conditions, up to 50% of hemocapillaries are nonfunctioning. The number of "open" capillaries depends on the intensity of the body. Blood flows through the capillaries at a rate of 0.5 mm/s under a pressure of 20–40 mm mercury.

Postcapillaries, or postcapillary venules, are vessels with a diameter of about 12–30 microns, formed by the merger of several capillaries. Postcapillaries have a larger diameter compared to capillaries. At the level of postcapillaries, active metabolic processes occur, and leukocyte migration is carried out.

The merger of postcapillaries forms venules. The collective venules are the initial link of the venular region of the MVB. They have a diameter of about 30–50 microns and do not contain smooth myocytes in the wall structure. Collective venules, the diameter of which reaches 50–100 microns continue into the muscle. In these venules, there are SMCs (the number of the latter increases with distance from the hemocapillaries) that are more often oriented along the vessel. In muscle venules, a clear three-layer wall structure is restored.

Unlike arterioles, there is no elastic membrane in the muscle venules, and the form of endothelial cells is more rounded. Venules remove blood from the capillaries, performing drainage and depositing (capacitive) function together with the veins. The reduction of longitudinally oriented smooth venous myocytes creates some negative pressure in their lumen, contributing to the "suction" of blood from postcapillaries. On the venous system along with blood from the organs and tissues are removed metabolic products.

The hemodynamic conditions in the venules and veins are significantly different from conditions in the arteries and arterioles due to the fact that the blood in the venous section flows at a low speed (1–2 mm/s) and at low pressure (about 10 mmHg).

Circulation

The cardiovascular system can be divided into two consecutively connected sections – a large (systemic) circulation and a small (pulmonary) circulation.

The systemic circulation begins with the left ventricle of the heart, throwing blood into the aorta. Numerous arteries depart from the aorta, and as a result, the bloodstream is distributed over several parallel regional vascular networks, each of which supplies a separate organ with blood. Arteries are divided dichotomously, and therefore, as the diameter of individual vessels decreases, their total number increases.

As a result of branching of the smallest arteries (arterioles), a capillary network is formed – a dense interweaving of small vessels with very thin walls. The total surface area of the capillaries in the body is huge – about 1000 m^2. Metabolic processes take place between the blood and cells in the capillaries. At the confluence of capillaries forms venules that are then collected in the veins. With merging, the number of vessels gradually decreases, and their diameter increases, and two veins approach the right atrium – the upper vena cava and the inferior vena cava.

In the abdominal cavity, blood flowing from the capillary networks of the mesenteric vessels and splenic vessels (i.e., from the intestines and spleen) in the liver passes through another system of capillaries, and only then goes to the heart. This is called portal circulation.

Through the arteries, blood flows to the organs, and through the veins, it flows from them.

The pulmonary circulation starts with the right ventricle of the heart, throwing blood into the pulmonary trunk. Then the blood enters the vascular system of the lungs, and then through the four large pulmonary veins flows to the left atrium and enters the left ventricle of the heart.

In the various sections of the vascular bed of the lungs, the arteries of the lungs and the veins of the lungs are much shorter, and their diameter is usually larger compared with the vessels of the corresponding sections of the systemic circulation. The walls of the large arteries of the lungs are relatively thin, while the small arteries of the lungs have thick walls with a developed muscular layer. There are no typical arterioles (i.e., resistance vessels) in the pulmonary circulation.

The diameter of the pulmonary capillaries is approximately 8 microns.

The Main Program Features of the Blood Supply System

So, the main task of CS is to ensure the implementation of all local programs, both social and "domestic" programs in the tissues and organs of the MO. To do this, the necessary components of the makers and objects must be delivered to specific sites for the execution of these programs, and the corresponding components produced in these organs and tissues for other local programs must be transported to the place of their demand.

When solving this basic task, the CS runs two local task programs.

The first of these, the convection duct of the necessary program components in the localization vicinity of all cells of tissues and organs, is performed by a complex maker consisting of an OM (the heart and large vessels) arteries and veins that together provide the blood pressure (BP) necessary for normal blood flow in all areas of the CS.

The vegetative motor innervation, humoral factors and cardiac automatization regulate the cardiac pressure program.

Vegetative innervation programs are carried out through the heart centers of the medulla oblongata and the pons. Impulses from the heart centers are transmitted through sympathetic and parasympathetic nerves, they relate to the frequency of contractions (chronotropic effect of the brain cardiac centers), contraction strength (inotropic effect of the brain cardiac centers) and velocity of the atrioventricular conduction (dromotropic effect brain heart centers). As in other organs, the transmitters of the nervous effects on the heart are mediators – acetylcholine in the parasympathetic NS and noradrenaline in the sympathetic NS.

Humoral factors affect cardiomyocytes that form the heart muscle, they have α-adrenergic, β-adrenergic, muscarinic acetylcholine receptors. Activation of α-adrenoreceptors helps to maintain the force of contraction. Agonists of β-adrenoreceptors cause an increase in the frequency and strength of contraction, muscarinic acetylcholine receptor causes a decrease in the frequency and strength of contraction. Noradrenaline is secreted from the axons of postganglionic sympathetic neurons and acts on the β1-adrenoreceptors of working atrial and ventricular cardiomyocytes, as well as the pacemaker cells of the sinus-atrial node.

Finally, the automatism of the heart ensures its contraction due to the action of impulses arising in a special part of the right atrium – a sinus-atrial node, called a heart rhythm driver, which is located at the confluence of the vena cava.

The second task program is to ensure the exchange of these program components between the blood and the intercellular interstitial fluid, the external exchange medium for the body's cells, performed by small vessels: arterioles, capillaries, venules, forming the so-called microcirculatory bed.

When executing both of these programs, the vessels, being composite, complex makers performing programs, simultaneously serve as sites for executing these programs.

The main cellular makers of both task programs are myocytes that make up muscle TMs and endotheliocytes that form the tissue of the endothelium lining the inner surface of all vessels.

Myocytes that are part of the muscle TM of the heart, while reducing it, perform the program of blood injection into the bloodstream.

In addition, myocytes form the middle muscular layer of the vascular wall, which is a maker that changes the lumen of the vessel in accordance with specific local programs running in this area of the CS.

In some places of vascular branching, for example, the place where performs the discharge of blood capillaries from the precapillary arterioles, myocytes form smooth muscle valve makers, sphincters that carry out programs for regulating blood flow to certain groups of blood vessels. In particular, in the absence of a pronounced functional load on the organ, most of the precapillary sphincters of this organ are closed.

Endotheliocytes are universal CMs, participating in the implementation of a number of both information and effector programs.

Such universality is ensured by the presence of a developed structure of social EMs in these cells, which in particular include various receptor molecules, makers synthesizing and secreting protein structures of participants in information and effector programs.

In the vessels of the microcirculatory part of the bloodstream in the peripheral zone of endothelial cells, there are different types of EM makers of transport programs: micropinocytosis vesicles, trans-endothelial channels, fenestrae and pores-hatches. Micropinocytosis vesicles provide for the transport of substances that are receptor-bound to the cell surface. Fenestrae are trans-endothelial channels with a diameter from 30 to 80 nm, closed by single-layer diaphragms. The pores are through-holes of the thinned areas of the endotheliocyte directly connecting the lumen of the vessel with the peri-endothelial space.

Endotheliocytes are involved in the implementation of both task programs of the CS.

They play crucial role in the regulation of vascular tone by supporting normal BP. They perform vasoconstriction, when it is necessary to restrict blood flow (e.g., in the cold to reduce heat loss), or their expansion in an actively working organ (muscle, pancreas during digestive enzymes, liver, brain, etc.), when it is necessary to increase its blood supply.

In response to the signal of receptors reacting to changes in blood flow velocity, endotheliocytes secrete factors (elemental makers) that regulate vascular smooth muscle tone: constrictors – endothelin, angiotensin II, thromboxane A2 and dilators – nitrogen oxide, prostacyclin, endothelial depolarization factor. As a result, the arteries continuously maintain their lumen at a level corresponding to the rate of blood flow through them, which ensures the stabilization of pressure in the arteries in the physiological range of changes in blood flow values.

The key role is played by the endothelium in the exchange programs carried out in the microvasculature.

As an example, we can trace the involvement of the endothelium in the auxiliary IP of leukocyte emigration to the area of localization of the attack of pathogens, the area of inflammation.

With the development of inflammation, the endothelial cells of the small vessels, capillaries and postcapillary venules quickly fall under the influence of pro-inflammatory factors, primarily cytokines.

Endothelial cells constitutively express receptors for these cytokines. Under the influence of cytokines, the endothelial cells themselves are activated.

The interaction of leukocyte adhesion molecules and activated endothelium cells serves as the basis for the program of leukocyte emigration from the bloodstream. In the initial situation, the interaction of leukocytes with cells of the vascular wall is fragile and the cells roll along the wall of the capillary or venule in the direction of blood flow.

Then there is the activation of leukocytes under the action of chemokines produced by endothelial cells, and due to local programs carried out by elementary leukocyte makers, integrins are activated.

As a result, the interaction of leukocyte integrins with receptors of endothelial cells becomes strong, and the cells stop.

After stopping, the leukocyte becomes available to the chemotactic signals supplied by chemokines and other chemotactic factors from the area of inflammation, to start a program of chemotactic transport to the area of inflammation.

In the advancement between endothelial cells, a crucial role belongs to homotypic interactions of two types, carried out by EM – molecules PeCAM (CD31) and CD99. PeCAM is involved in the entry and exit of lymphocyte into the interendothelial space, and CD99 – in overcoming the contact zone between endothelial cells.

The structure contributes to the effective participation of endotheliocytes in the implementation of the task program for the exchange of program components between the blood and intercellular body tissue fluid. So, the luminal (facing the vessel lumen) surface of the endotheliocyte includes three layers: glycocalyx, plasmalemma and submembrane (cortical).

This ensures the reception and selection of the portable components, the regulation of the transport properties of the endotheliocyte, determines the necessary changes in the configuration of the cell surface, in particular, forms micro-growths, folds and microvilli that contribute to the capture of material from the lumen of the vessel.

The lateral surfaces (contact area) are filled with a cementing substance of the para-plasmalemma layer and contain specialized intercellular contacts that can be simple, complex, dense and slit. Contacts are dynamic structures capable, due to remodeling of the cytoskeleton, to change the size of the intercellular gaps within minutes, adapting to the conditions for the execution of border transport programs based on receptor signals of endothelial cells.

In addition to the implementation of the main task-oriented transport programs, endotheliocytes are involved in the implementation of such important additional programs as the regulation of coagulant programs performed by CMs with platelets, and angiogenesis, which is the formation of new blood vessels.

MANAGEMENT PROGRAMS

As noted above, a MO is a social formation of specialized cells, therefore, it inherits the basic program and information principles implemented in the cell. This also applies to MO control programs that are based on a signal chain, in which start is initiated by the primary control signal, and ending with an information impact on the controlled maker, most often elementary or cellular. We have considered these chains in Chapter 1. They include amplification and signal conversion links, as well as intelligent links for analyzing the signals received by them and making decisions on the subsequent sending of output signals to specific makers.

The distance between the source of the control primary signal and the controlled maker, i.e., chain length varies in MO in a very wide range from cell size to the size of the entire MO. The required signal transit time varies within large limits, from fast response speeds when performing control of motion programs to hormonal signals of slow reorganization of many program subsystems.

For the implementation of these chains in MO, there are two information control program subsystems – fast nervous and slow humoral. The basis of the NS consists of nerve cells, in which two mechanisms, electrical and chemical, are used to carry out the signal transfer program. In the humoral system, signal transmission is carried out by transporting the control information molecule to a controlled maker by a convection manner in body fluids composed of blood and lymph and by diffusion in the intercellular fluid.

Humoral System

Humoral control programs (HCPs) were formed at the early stages of the evolution of MOs and are carried out through the body's fluid media (blood, lymph, tissue fluid, saliva) using, to a large extent, hormones secreted by cells, organs, tissues. In highly developed animals, including humans, the HCP is subject to the NS and together with it makes up a unified system of control programs MO.

Humoral programs are carried out by transmitting signals using various biologically active substances, including hormones, neurotransmitters, prostaglandins, cytokines, growth factors, endothelium, nitrogen oxide and a number of other substances through the body fluids.

Humoral programs are divided into two groups: endocrine and local.

Endocrine programs are information programs that release hormones into the bloodstream system and are implemented by cellular and OMs, the endocrine glands. Hormones are biologically active substances that are transported by the convection transport programs of the circulatory system and have specific regulatory effects on the vital activity of cells and tissues. Hormones are delivered to almost all cells of the MO, but the response to the action of the hormone can only be from those cells (targets), on the membranes in the cytosol or nucleus of which there are receptors for the corresponding hormone.

A distinctive feature of the local HCPs is the fact that biologically active substances produced by the CM do not enter the bloodstream but influence the cell producing them and its immediate environment, spreading due to diffusion through the intercellular fluid. Such control programs are divided into autocrine, paracrine and juxtracrine.

When performing an autocrine program, the signaling molecule enters through the cell membrane of the cell information maker into the extracellular fluid and binds to the receptor on the outer surface of the membrane. Thus, the cell reacts to a signaling molecule synthesized in it – the ligand. The attachment of the ligand to the receptor on the membrane causes the activation of this receptor, and it launches a program of biochemical reactions in the cell, which provide a change in its functioning. This is particularly necessary to maintain a stable level of secretion of certain hormones. For example, in preventing the excessive secretion of insulin by B cells of the pancreas, the inhibitory effect of the hormone secreted by them has an important role.

The paracrine program is carried out by the secretion of cell signaling molecules that go into the intercellular fluid and affect the vital activity of neighboring cells. A distinctive feature of this program is that in the transmission of a signal, there is a stage of diffusion of the ligand molecule through the intercellular fluid from one cell to another neighboring cells. Thus, insulin-secreting pancreatic cells affect the gland cells secreting another hormone – glucagon.

Finally, the juxtracrine program is carried out by transferring signal molecules directly from the outer surface of the membrane of one cell to the membrane of another. This occurs under the direct contact condition of membranes of two cells. Such attachment occurs, for example, when leukocytes and platelets interact with the endothelium of blood capillaries in a place where there is an inflammatory process. On the membranes lining the capillaries of cells, signal molecules appear at the site of inflammation, which bind to receptors of certain types of white blood cells. Such binding leads to the activation of the attachment of leukocytes to the surface of a blood vessel. This can be followed by a program that ensures the transfer of leukocytes from the capillary to the tissue and their suppression of the inflammatory reaction, which is described in the IPs.

Nervous System (NS)

Neurons Form NS Signal Chains

The main CM of NS is a neuron. The different types of neurons form the signal chains. Each neuron consists of a cell body containing a nucleus and processes diverging from it. Depending on their number, they are distinguished as unipolar (with one process), bipolar (with two processes) and multipolar (with multiple processes) neurons.

One of the processes is called an axon. It is longer than the others and it does not branch like a tree, but lateral processes (collaterals) move away from it at a right angle. At the end, the axon branches into small twigs (terminals). The number of terminals varies from units to several thousand.

Another type of nerve cell process is dendrites. There may be several of them. The dendrites are much shorter than the axon. They branch dichotomously, and branches diverge at acute angles. They also have several ways of branching.

The main program executed by a neuron is the transfer of a signal, a nerve impulse from a dendrite through the body of a neuron to an axon.

In the input sensory links of the information signal chains, these signals contain information about stimuli of a certain type.

In effector output links, the signals are commands for controlled makers – most often muscles, so the neurons of the output links are also called motors.

Finally, in interneurons (intercalary neurons) connecting one neuron with another, the signals provide a complex interaction and combining information from several different sources.

Signal Transfer Program

Signal transfer by the nerve cell is carried out by changing the local electric potential of the plasma membrane of the neuron in the direction from the dendrite to the axon.

At rest, the membrane potential of a neuron has the same negative value everywhere – the internal environment of the cell is electronegative with respect to the extracellular medium. The potential difference depends on significant gradients of the concentrations of Na^+ and K^+ created by the Na^+ -K^+ pump. Due to the leakage channels, the K^+ membrane at rest is permeable only to potassium; therefore, the resting membrane potential is close to the equilibrium potassium potential – usually around 70 mV.

Electrical disturbance in one part of the cell spreads to other areas. In this case, the electrical signal may take the form of depolarization when the potential drop on the membrane decreases, or hyperpolarization, at which it increases.

This disturbance fades as it moves away from its source if there is no additional amplification along the signal path. At short distances, the attenuation is insignificant, and many small neurons pass signals passively, without amplification in accordance with their cable properties. Such cells often do not have or almost do not have potential-dependent Na^+ channels that are necessary for the action of an active mechanism of signal propagation, accompanied by its amplification. Such an amplified signal is called an action potential.

The action potential occurs when the membrane instantly depolarizes to a level above a certain threshold. As a result of such depolarization of the membrane section, potential-dependent sodium channels will open here, which will cause a current of Na^+ ions along their electrochemical gradient. The consequence will be further depolarization of the membrane, resulting in an even greater number of Na^+ channels opening up, and so on, like a chain reaction, until the potential in this section of the membrane approaches the sodium equilibrium potential.

Due to the cable properties of the axon, the local influx of a large number of Na^+ ions during an action potential leads to the emergence of longitudinal currents that depolarize adjacent portions of the membrane to a threshold level, which in turn causes the occurrence of action potentials. This process spreads along the axon from one excited area to another at a speed that in vertebrates can range from 1 to 100 m/s, depending on the type of axon.

The speed of the pulse depends on the cable properties of the axon and is achieved by isolating the surface of many axons with the myelin sheath, which drastically reduces the capacity of the axon's membrane and at the same time almost completely prevents leakage of current through it.

The myelin sheath is formed by specialized glial cells – Schwann cells in the peripheral and oligodendrocytes in the central NS. The plasma membrane of these cells is wound onto the axon, layer by layer tightly in a spiral. Each Schwann cell myelinates one axon, forming a shell segment of about 1 mm in length containing up to 300 concentric layers; oligodendrocytes form similar segments of the shell at the same time in several axons.

Between the two adjacent segments of myelin there remains a narrow unprotected section of the membrane, in which almost all the sodium channels of the axon are concentrated, the so-called Ranvier traps with a width of only about 0.5 mkm, whereas in the areas covered by the myelin sheath, they are almost absent.

Therefore, the currents associated with the action potential in the intercept area are effectively guided by passive conduction to the next intercept, quickly depolarize the membrane and excite the next action potential. Such conduction is called the saltation when the signal spreads along the axon, "jumping" from one interception to another.

Returning the Neuron to Its Original State after a Signal Transfer Program

The transition of the neuron to the initial state of rest with the initial negative value of the membrane potential occurs due to two factors.

First, the Na^+ channels spontaneously go into a closed, inactivated state.

Second, potential-dependent K^+ channels open. These potassium channels react to changes in membrane potential in almost the same way as sodium, but less quickly, and therefore they are sometimes called slow K^+ channels.

As soon as the K^+ channels open, the outgoing potassium current quickly overlaps the incoming sodium current in size and the membrane potential returns to the level of the equilibrium K^+ – potential even before complete inactivation of Na^+ channels. As a result of repolarization, the potential-dependent potassium channels are closed again, and inactivated sodium channels are transferred to the original closed state, but capable of activation. Thus, the ability to generate action potentials can recover in a given area of the membrane in less than one thousandth of a second.

Program of Signal Transmission between Neurons

A synapse serves to transmit a nerve impulse between two cells. A synapse is the point of contact between two neurons or between a neuron and an effector cell that receives a signal. During synaptic transmission, the amplitude and frequency of the signal can be adjusted. The transmission of pulses is carried out chemically by means of mediators or electrically.

Transmission is usually indirect. The cells are electrically isolated from each other: the presynaptic cell is separated from the postsynaptic one by the synaptic cleft. A change in the electrical potential in a presynaptic cell leads to the release of a substance called a neurotransmitter, which is stored in membrane-bound synaptic vesicles and released by exocytosis. The neurotransmitter diffuses through the synaptic cleft and with the help of receptors causes a change in the electrophysiological state of the synapse of the dendrite of the postsynaptic neuron, i.e., appearance of an electrical signal to be transferred by this neuron.

There are neuron receptors linked to the channel or not.

Channel-linked receptors are actually ligand-dependent channels. The conformation of such receptors immediately after the binding of the neurotransmitter changes in such a way that an open channel for certain ions forms in the membrane and, as a result, the permeability and potential of the membrane change. Receptors of this type serve as the basis for the most common and most studied method of signal transmission in chemical synapses, in which transmission occurs very quickly.

Receptors that are not linked to the channel trigger the same processes as when exposed to water-soluble hormones and local chemical mediators throughout the body. In such receptors, the neurotransmitter-binding sites are functionally linked to an enzyme that, in the presence of a neurotransmitter, usually catalyzes the formation of an intracellular mediator, for example, cAMP. In turn, this mediator causes changes in the postsynaptic cell, including modification of ion channels in the cell membrane. Unlike the channel-linked receptors, these receptors, as a rule, mediate the relatively slowed, but more prolonged effects of neurotransmitters. Activation of such receptors is believed to cause changes in neurons that persist for a long time and underlie learning and memory.

Program to Restore Neuron Activity

To restore the biochemical program activity of the neuron and, in particular, its synapses, it is necessary to implement programs for updating the pool of EMs in the cell body and their subsequent transport along the dendrites and axons in the direction of the synapses. Direct transport – from the cell body along the processes to their periphery transfers secreted proteins and membrane-bound molecules to synapses.

There is also a reverse (retrograde) transport – along the processes of the neuron to the cell body, which carries, in particular, membrane vesicles, spent organoids and their components.

The transport programs in neurons are carried out by EMs – the so-called motor molecules of the kinesin, dynein and myosin proteins. The motor part of such molecules binds to the microtubule of the cell, and its tail part to the transported material. When energy released, for example, during the hydrolysis of ATP molecules, is absorbed by the transport protein – maker allosteric changes occur in it. This leads to its movement along the microtubule along with the transported material. A number of auxiliary makers, adapter proteins associated with transport makers, also take part in providing transport along the processes.

Features of Neural Chain Links

Let us consider some features of the programs executed in various links of neural chains.

Input Sensor Links

In the input links, signals are generated from the sensors – the main makers of these input neurons.

Any signal received by the NS is converted to the electrical signal. All sensory cells are converters of various input signals – chemical, temperature, sound, optical and others into electrical impulses. In some sensory organs, the transducer is part of a sensory neuron conducting impulses, while in others it is part of a sensory cell that is specially adapted for signal conversion but not involved in long-distance communication. Such a cell transmits its signals to its associated neuron through a synapse. But in both cases, the effect of an external stimulus causes an electrical shift in the transducer cell, called a receptor potential, which is similar to the postsynaptic potential (PSP) and also ultimately serves to regulate the release of the neurotransmitter from another part of the cell. Just as in the synapse, external stimuli can affect the electrical state of the cell both directly, acting on the ion channels, and indirectly through receptor molecules that trigger the synthesis of the intracellular mediator that already affects the ion channels. It is believed that the auditory sensory cells use a direct mechanism of action involving receptors associated with the channels, and the sensory cells of the eye use an indirect path through receptors associated with the G-protein.

The ear perceives not only sound signals, but also information about the direction of gravity and accelerated motion. The basis of all the sensory functions of the ear is mechanoreception – the registration of small displacements of the environment surrounding the sensory cells located in the inner ear. On the upper surface of such a cell is a bunch of giant microvilli, called stereocilia, which are in this case the receptor EM.

Such cells are called hair cells that, as cell receptor makers, perform programs for recording accelerations and gravity and sound signals.

When bending stereocilia in hair cells, receptor potential arises; in the case of linear acceleration and gravity registration programs, bending of stereocilia occurs due to the displacement of the gelatinous extracellular matrix hanging over the hair cells and connected to the tips of stereocilia.

Auditory hair cells performing a program for detecting sound signals are arranged in rows on the basilar membrane – a narrow and long elastic partition between two fluid-filled spiral channels that run parallel to a specific part of the inner ear called the cochlea. Sound waves cause the eardrum to vibrate, and through the bones of the middle ear, this vibration is transmitted to the fluid in the cochlea channels and further to the basilar membrane, the vibrations of which cause the stereocilia of the auditory hair cells to tilt. Due to the special structure of the cochlea, different parts of which resonate to varying degrees depending on the frequency of the sound waves, the distribution of the activated hair cells provides information about the pitch of the sound.

Two types of receptor CMs perform programs for recording light signals: cones and rods.

Cones are used for color vision and perception of small details and require a relatively strong light. Rods provide monochromatic vision in low light and can give a measurable electrical response to a single photon. The mechanism of action of rods and cones is similar, but the rods are better understood.

The rod consists of an outer segment containing a light-perceiving apparatus, an inner segment where there is a multitude of mitochondria, a nuclear region and a synaptic body that forms contact with the retinal nerve cells.

The outer segment, where the key stages of the light signal transformation occur, is a cylinder containing about a thousand discs tightly packed in a stack. Each disk is formed by a membrane closed in a bubble, in which there are photosensitive molecules of rhodopsin – the primary receptor EM. When a photon is absorbed, the rhodopsin molecule changes its conformation, which causes the implementation of a cascade program of hydrolysis of the cGMP information maker. This in turn leads to the closure of sodium channels in the plasma membrane of the rod.

In the dark, these sodium channels are open, which maintains the initial potential of the cell in a state of depolarization. In this case, the voltage-dependent calcium channels of the synaptic body are maintained in the open state, and the passage of Ca^{2+} ions into the cell leads to the continuous release of the mediator in the synaptic cleft.

When exposed to light, the closure of sodium channels results in hyperpolarization of the rod, the influx of Ca^{2+} ions decreases and, accordingly, the release rate of the mediator decreases. Since the mediator has an inhibitory effect on many of the postsynaptic neurons, these neurons are disinhibited when illuminated and as a result are excited.

Signal Processing Programs Run by Interneurons and Effector Neurons

Let us consider some examples of programs for processing and combining signals from different sources, carried out, as a rule, in CMs – interneurons, although often such programs are performed by directly effector links of nerve chains, for example, motor neurons.

For example, on a typical spinal cord motor neuron, synapses form thousands of nerve endings from hundreds and possibly thousands of different neurons; the bodies of the neuron and dendrites are almost completely covered by synapses.

Some of these synapses transmit signals from the brain, others deliver sensory information from the muscles and skin, and report the results of "calculations" produced by intercalary neurons of the spinal cord. A motoneuron must integrate information from these multiple sources and either respond by sending signals along an axon or remain at rest.

Of the many synapses on the motoneuron, some will tend to excite it, and others to inhibit. Although all endings of the axon of a given motor neuron emit the same neurotransmitter, the motor neuron has many different receptor proteins concentrated on different postsynaptic sites of its surface. In each of these areas, under the action of presynaptic pulses, a certain group of channels is opened or closed, as a result of which a characteristic potential change occurs in the motoneuron – a PSP arises. Depolarization corresponds to excitatory PSP (arising, e.g., when opening channels for sodium ions), and hyperpolarization corresponds to inhibitory PSP (appearing, e.g., when opening chloride channels).

Although the dendrite membrane and the bodies of most neurons are rich in receptor proteins, it contains very few potential-dependent sodium channels and is therefore relatively non-excitable. Single PSPs, as a rule, do not lead to the emergence of action potential. Each incoming signal is displayed by the magnitude of the local gradual PSP that decreases with distance from the input synapse. If signals simultaneously arrive at synapses located in the same area of the dendritic tree, then the total PSP will be close to the sum of the individual PSPs, and the inhibitory PSP will be counted with a negative sign. At the same time, the total electrical perturbation that occurred in one postsynaptic region will spread to other regions due to the passive cable properties of the dendrite membrane.

The body of the neuron, where all the effects of the PSP converge, is usually small (less than 100 microns in diameter) compared to a dendritic tree, the length of the branches of which can be measured in millimeters. Therefore, the membrane potential of the cell body and the parts of the processes closest to it will be approximately the same – this will be the cumulative result of the effects produced by all input signals, taking into account the distance of one or another synapse from the neuron body. Thus, it can be said that the total PSP of the cell body is the result of the spatial summation of all received stimuli. If excitatory input signals dominate, then the cell body is depolarized, if the inhibitory – usually hyperpolarized.

The temporary summation combines the signals received at different times.

The neurotransmitter released after the action potential arrives at the synapse creates a PSP on the postsynaptic membrane, which quickly reaches its peak (due to the short-term opening of ligand-dependent ion channels) and then exponentially (determined by the membrane capacity) decreases to its initial level. If the second impulse comes before the first PSP is completely attenuated, then this second PSP is added to the remaining "tail" of the first. When, after some period of rest, a long volley of rapidly repetitive pulses arrives, each successive PSP will overlap with the previous one, resulting in a large PSP, the magnitude of which reflects the frequency of the discharge of the presynaptic neuron. Thus, the essence of time summation is that the frequency of the received signals is converted into the value of the total PSP.

Due to temporal and spatial summation, the membrane potential of the body of one postsynaptic neuron is regulated by the frequency of discharges of many presynaptic neurons.

As a result of the integration of all input signals, the postsynaptic cell forms a definite response, usually in the form of pulses, to transmit signals to other cells, often in distant parts of the body. This response signal reflects the magnitude of the total PSP in the cell body. However, although the total PSP smoothly changes all the time, action potentials have a constant amplitude and obey the "all or nothing" rule.

The only variable when transmitting signals using pulses is the time interval between successive pulses.

Therefore, to transmit information over long distances, the value of the total PSP must be transformed, or recoded, into the frequency of a pulsed discharge. Such coding is achieved using a special group of potential-dependent ion channels concentrated at the base of the axon in an area called the axon hillock.

The conduction of nerve impulses depends mainly on potential-dependent sodium channels. Initially, the axon hillock membrane generates the impulses, where there are many of such channels. But for the implementation of a special coding function, the axon hillock membrane must contain at least four more classes of ion channels – three selectively permeable for potassium ions and one permeable for Ca^{2+}. Three types of potassium channels have different properties and are called slow, fast and Ca^{2+}-dependent potassium channels by their specific actions.

The considered programs for processing and combining signals, carried out in the body of neurons – motoneurons, and more often interneurons, allow to realize the whole complex of operations necessary to control the target postsynaptic CMs, which in turn can be next-order interneurons, motoneurons or effectors-muscle makers or secretory cells.

Taking into account the noted possibility of realization of two polar effects on a neuron in synapses – excitation and inhibition when combining signals on a neuron, all basic discrete logic operations can be implemented: conjunctions, disjunction, inversion, implication.

In addition, analog arithmetic operations of summation and subtraction can be used to estimate the net impact on a neuron.

Effector Neurons

The main effector makers, controlled by the neural signal chains of a MO, are muscle and secretory CMs. Most of the muscle makers of the MO are SMCs that are controlled by sympathetic (adrenergic) and partly parasympathetic (cholinergic) effector neurons. Neurotransmitters diffuse from the varicose terminal extensions of such neurons into the intercellular space. Subsequent interaction of neurotransmitters with their receptors in the plasmalemma causes a contraction or relaxation of the SMC. It is significant that in the composition of many smooth muscles, as a rule, not all the SMCs are innervated. Excitation of SMCs without innervation occurs in two ways: to a lesser extent with slow diffusion of neurotransmitters and to a greater extent through gap junctions between the SMC.

For example, stomach makers are controlled from two sources of efferent innervation – two program information channels: the parasympathetic (from the vagus nerve) and the sympathetic (from the border sympathetic trunk). Three nerve plexuses are located in the wall of the stomach: intermuscular, submucous and subserous.

The excitation of the vagus nerve accelerates the reduction of the stomach and increases the excretion of GJ by the glands. The excitation of sympathetic nerves, on the contrary, causes a slowdown of the contractile activity of the stomach and a weakening of the gastric secretion.

Network Structures Formed by Neurons

So, we discovered that, inheriting the chain-based management structure of single-cell, informational makers of MO also perform three types of local control information programs: initial, starting – receptor or sensory, final, finishing – effector and intermediate – summing, processing, decisive.

At the same time, the standard task management program from these local programs is implemented.

In this case, the standard task management program is collected, which is collected from these local programs.

First, receptor, sensory information makers collect the primary signaling situation corresponding to this task program.

Then, this primary signal information is processing and summing up by intelligent makers until the information situation is obtained, which is necessary and sufficient to make a decision on the launch of the final effector local program.

And, finally, the command neuron (CN) maker launches this effector program.

In complex MOs of animals and humans, a separate cellular information maker – a neuron is unable to control the MO in any way. All the basic forms of the NS's activity are associated with the participation in its programs of certain groups of nerve cells – the so-called network structures, i.e., composite information makers.

It should be noted that the simplified idea of a linear multi-link signal chain with single-channel input and output of the signal in links is rarely implemented in practice. A typical situation corresponds to the case of multi-channel entry into the links and single-channel exit from them. The most common type of neuron has a combination of dendrites and one axon.

Through dendritic channels, the operational neuron (ON) receives signals from external and internal body sensors about the information situation that corresponds to or does not correspond to the output effective action potential to the ON axon.

In other words, ON, processing the input signals, makes a decision about whether or not to output the signal. To put it simply, we can say that ON produces a signal of an action potential when operations to combine input signals give a result confirming the maturation of the information situation to the level necessary and sufficient for the decision to generate a reaction signal.

Primary situational input dendritic signals may come from sensory cells, for example, the eyes. Certain groups of cells in the visual centers generate action potentials when the eye sees a line oriented in space. The output signals from these groups will be received by other neurons that perform the next stage of the recognition program of the object being observed, and so on, to ever higher levels of analysis of this object until the last ON in this chain decides on the final recognition to generate an action signal for the motor neuron launching a program of muscle effector makers, for example, on capturing an object.

In the considered example, the complex network information maker consists of sensory visual neurons, analyzing optical information of different types to the level of recreating the situation necessary and sufficient to launch the final effector part of the task program to be solved and, finally, the CNs makers launching this final program stage.

Traditionally, the above-described example of integration of nerve cells into neural networks is denoted by the term convergence, when the same nerve cell can receive impulses from different receptors of the body, i.e., signals of a wide variety of stimuli.

The neurons of these departments receive information from the receptors of relatively small areas of the body – the receptive fields of the same reflex. In the suprasegmental regions, especially in the cerebral cortex, impulses of various origins from different reflex paths converge. The neurons of the suprasegmental departments can receive signals about light, sound, proprioceptive and other stimuli, i.e., signals of different modality. "Convergent patterns" are constantly changing on the body of neurons – excited and inhibited parts. It is estimated that the sizes of the receptive fields of cortical neurons, that is, the areas of the body from which afferent stimuli can enter them, are 16–100 times larger than the dimensions of the same fields for afferent cells of spinal reflex arcs. Due to this diversity of incoming information, wide interaction, matching, selection, development of adequate reactions and establishment of new connections between reflexes can occur in the neurons of the overlying brain regions.

However, the formation of local neural networks is possible with the use of another mechanism, the so-called divergence, when one neuron is in contact with many higher order neurons and sends signal to different segments of the spinal cord and the brain.

The divergence of the signal path is also observed in intercalated, for example, command, cells, as well as output nerve cells (motoneurons and other effector neurons). Thus, in humans, one large motor neuron (group A) innervates, excites dozens of muscle fibers (in the external eye muscles) and even thousands (in the muscles of the limbs).

Local networks occur in the form of various structures of the NS.

For example, ganglia, nerve nodes are formed most often in the digestive tube MO.

Somatosensory nuclei are in the posterior horns of the gray matter of the spinal cord, somatomotor nuclei are in the anterior ones and vegetative nuclei are in the lateral horns. The functional significance of the nuclei of the spinal cord at the level of each of its segments has been established. In the process of embryogenesis, the somatomotor nuclei of the spinal cord develop from the main plate of the neural tube.

In the brain stem, the nuclei of the cranial nerves, the nuclei of the reticular formation and their own nuclei are isolated. In most nuclei of the brain stem, cranial nerves begin or end. The location of the nuclei of the cranial nerves in the brain stem is different from the position of the nuclei in the spinal cord. Two rows of somatomotor nuclei are formed near the median line in the brain stem: the dorsal, represented by the nuclei of the motor cranial nerves, and the ventral, represented by the motor nuclei of the mixed cranial nerves.

The nerve center is the set of nerve cells needed to perform a function. These centers respond with appropriate reflex responses to external stimulation from receptors associated with them. The cells of the nerve centers also react to their direct irritation by substances in the blood flowing through them (humoral effects).

Complex reactions in the whole organism are usually associated with the participation of many nerve centers located on different floors of the central NS. For example, an arbitrary change in breathing is carried out by a person with the participation of the centers of the cerebral cortex, a special center in the midbrain, the respiratory center, the medulla oblongata and the centers of the spinal cord that innervate the respiratory muscles.

When several local networks are combined, nerve centers are formed, which are a complex of elements necessary and sufficient to carry out a certain reflex or a more complex behavioral act. In turn, the nerve centers located in different parts of the brain can cooperate in the so-called distributed systems that coordinate the activities of the organism as a whole. These systems have a hierarchical structure and represent the next, higher integrative level of the central NS, whose activity is based on some general principles of the nerve centers.

The considered examples of local neural networks are formed to solve specific information tasks implemented by all types of makers: sensory, intellectual (processing, summing) and command.

Such networks form the basis of both the peripheral and central parts of the NS.

Functionally, the NS forms two sections: somatic and autonomous (vegetative).

The somatic department regulates MO behavior in the external environment and is associated with the work of skeletal muscles mainly.

The vegetative department regulates the work of smooth muscles, internal organs, blood vessels and consists of two subdivisions – the sympathetic and the parasympathetic that act on the principle of complementarity.

Consider an example of a local neural network in the peripheral part of the NS. It is known that many internal organs, extracted from the organism, continue to realize their inherent functions. For example, the peristaltic and absorption programs are performed in the intestinal digestive tract, etc. Such relative functional independence is explained by the presence in the walls of these organs of the meta-sympathetic part of the autonomic NS, which has its own neurogenic rhythm and full set of independent reflex activity links – sensory, associative, effector with the corresponding mediator provision.

This system contains its own sensory elements (mechanical, chemo-, thermo-, osmoreceptors) that send information about the state of the innervated organ to their internal networks and are also capable of transmitting signals to the central NS. The sphere of innervation of the meta-sympathetic part of the autonomic NS is limited and covers only the internal organs. For these organs, the meta-sympathetic innervation is basic. The efferent connections of the meta-sympathetic part with the central structures are mediated by the neurons of the sympathetic and parasympathetic parts of the autonomic NS, forming synaptic contacts on the bodies and processes of the meta-sympathetic interneurons and effector neurons.

The organs of the central NS, which form the spinal cord and the brain, in fact also consist of neural networks of a higher hierarchical level, to which, in particular, the nerve nuclei and centers have already been mentioned.

The brain is in the skull. The bodies of brain neurons are located in the gray matter of the cortex and nuclei scattered among the white matter of the brain. White matter consists of nerve fibers connecting various centers of the brain and spinal cord.

Spinal cord is located in the spinal canal. At the top, the spinal cord passes into the head, at the bottom ends at the level of the second lumbar vertebra, a bundle of nerves extending from it resembling a horse's tail.

The bodies of spinal cord neurons are concentrated in gray pillars that occupy the central part of the spinal cord and stretch along the entire spine.

There are ascending nerve pathways, along which nerve impulses go to the brain, and descending nerve pathways, along which excitation goes from the brain to the centers of the spinal cord.

The spinal cord performs reflex and conductive functions.

The centers of the spinal cord work under the control of the brain. The impulses from it stimulate the activity of the centers of the spinal cord and maintain their tone. If the connection between the spinal cord and the brain is broken, which happens when a spinal cord is damaged, shock occurs. In shock, all reflexes whose centers lie below the damage to the spinal cord disappear and voluntary movements become impossible.

Spinal nerves exit the spinal canal through the intervertebral foramina, then their motor fibers are sent to the muscles, and the sensory fibers go to their endings in the skin, muscles, joints and internal organs. The connection of each segment with the area of innervation is carried out according to a rigid topographic scheme: the motor fibers control strictly defined muscles, and the sensitive ones receive information from certain regions (in the skin these are limited areas or dermatomes).

In the spinal cord distinguish between gray and white matter. The centrally located gray matter is dominated by the bodies of nerve cells, whereas the white matter consists mainly of many processes of neurons: information passes through them from one spinal cord segments to another, from the spinal cord to the brain (ascending paths) and vice versa, from the brain to the spinal (descending path).

The spinal cord is phylogenetically the oldest structure of the brain and most of the neural connections in it are very stable, the functionally different neurons are perfectly matched to each other. This allows the spinal cord to independently regulate the simplest motor and autonomic reactions, such as detaching the hand from a hot object or emptying the bladder with a significant stretching of its walls. But even with such standard reactions, the spinal cord is under constant control of the brain. The spinal cord supplies sensory information to it, and from it receives most of the motor programs and instructions on the part of the autonomic regulation.

Rostrally from the spinal cord is the medulla, its direct continuation is the bridge, delimited by a sharply defined protrusion – it is formed by numerous fibers that serve to communicate with the cerebellum. The midbrain is located rostrally from the bridge.

Motor and sensory neurons of the brain stem are only a small part of its gray matter. Most stem neurons specialize in information processing, their clusters form numerous nuclei, the processes of which can be directed into the spinal cord, forming downward paths or linking the trunk with other regions of the brain.

In the medial part of the trunk throughout its length contains a diffuse network of neurons that form the reticular formation. Numerous branching processes of its neurons receive afferent information from all sensory systems whose conductors pass through the trunk. The reticular formation integrates sensory signals and, in accordance with their nature, affects the activity of the brain and spinal cord. The reticular formation of the brain mainly has an activating effect; its downward influence can be both activating and inhibiting. Some nuclei of the reticular formation perform narrowly specific functions, such as, for example, regulation of arterial pressure or control of skeletal muscle tone. It plays a very important role in regulating the sleep cycle and attention.

The cerebellum receives sensory information from all systems that are excited during movement, as well as from other regions of the brain that are involved in the creation of motor programs. Information is transmitted to the cerebellum and from it via the numerous nerve fibers that form the legs of the cerebellum: three pairs of such legs anatomically and functionally connect the cerebellum with the brain stem.

The structure of the cerebellum is quite complex: it has its own cortex, consisting of a huge number of cells of several varieties, and under the cortex, among the white matter of conducting fibers, are several pairs of nuclei of the cerebellum. The function of the cerebellum first consists in the formation of motor programs necessary for maintaining balance, regulation of strength and volume of movements. The role of the cerebellum in the regulation of fast movements is especially important.

The two important adjacent structures of the brain are thalamus and hypothalamus. Thalamus is located on both sides of the third ventricle of the brain and contains a large number of switching nuclei. The thalamus is an extremely important center for processing almost all sensory information; it is the main switching station on the way to transmit information to the cortex. Some thalamic nuclei receive sensory information from the periphery, process it and transmit to certain topographic areas of the cortex, which specialize in analyzing only one type of information – visual, auditory, somatosensory (perceiving signals from the surface of the body and from skeletal muscles). Thalamic nuclei of this type are called specific or projective. The nuclei of a different type, nonspecific, receive signals mainly from neurons of the reticular formation; such information does not carry information about the specific qualities of stimuli acting on the body. The neurons of the nonspecific nuclei of the thalamus affect different areas of the cortex. In turn, the neurons of the cerebral cortex can influence the activity of the thalamic neurons. The connection between the thalamus and the cortex can be called bilateral.

Along with sensory there are motor nuclei in the thalamus. Connections between the motor cortex, cerebellum and subcortical nuclei are established with the help of neurons of these nuclei when forming motor programs. Another group of nuclei of the thalamus is necessary in order to ensure the interaction of different regions of the cortex with each other and with subcortical structures. Such nuclei can be called associative. Due to its numerous connections with different regions of the brain, the thalamus is involved in the implementation of many functions, for example, with his participation the limbic system forms emotions and the hypothalamus controls the work of the internal organs.

The hypothalamus is located in the ventral part of the diencephalon directly above the pituitary gland. It is the highest center of regulation of the vegetative functions and coordinates the activity of the sympathetic and parasympathetic divisions of the autonomic NS and coordinates it with the motor activity. It also controls the secretion of pituitary hormones, thereby controlling the endocrine regulation of internal processes. Some of the numerous nuclei of the hypothalamus regulate the body's water-salt balance, body temperature, hunger and satiety and sexual behavior. The hypothalamus is the most important motivational structure of the brain, in this regard, it is directly related to the formation of emotions and to the organization of purposeful behavior. The functions of the hypothalamus are provided through its bilateral relations with many regions of the brain and with the spinal cord. In addition, many hypothalamic neurons are able to directly respond to changes in the internal environment of the body.

Symmetrically located hemispheres of the brain are connected to each other by approximately 200 million nerve fibers, forming the so-called corpus callosum. In each hemisphere, the cerebral cortex and subcortical nuclei located deep in the hemispheres are distinguished: the basal ganglia, the hippocampus and the tonsils of the brain.

The basal ganglia receive input from all areas of the cortex and the brain stem, and through the nuclei of the thalamus and from the cerebellum and use it to form motor programs. In addition, the basal ganglia are involved in the cognitive activity of the brain.

The hippocampus and tonsils are important components of the limbic system of the brain that forms emotions. The hippocampus is also needed to transform short-term memory into long-term memory. Tonsils coordinate vegetative and endocrine reactions associated with emotional experiences.

The cortex represents the outer surface of the hemispheres, it is the largest region of the brain by the number of cells. The thickness of the gray matter of the cerebral cortex varies between 1.5 and 5 mm, neurons are located in it in layers. In the greater part of the cortex, there are six layers, differing in composition of the cells forming each layer.

Depending on the functions performed, different areas of the cortex are divided into sensory, motor and associative areas. The sensory areas include the somatosensory cortex, which occupies the posterior central gyri; the visual cortex, which is located in the occipital lobes and the auditory cortex, which occupies part of the temporal lobes. The motor cortex is in the anterior central gyrus and in the regions adjacent to gyrus of the frontal lobes. The associative cortex occupies the rest of the brain surface and is divided into the prefrontal cortex of the frontal lobes, parietal-temporal-occipital and limbic. The prefrontal cortex creates plans for a complex of motor actions, the parietal-temporal-occipital integrates all sensory information and the limbic one participates in the formation of memory and emotions and determines the motivational aspects of behavior.

So, in accordance with the modern view on the functional structure of the NS, it consists of local neural networks formed in its peripheral and central parts, in accordance with the tasks that they solve – from simple reflex to complex behavioral. At the same time, networks performing complex behavioral programs consist of a set of simple and complex sensory, effector and analytical networks of various levels of complexity and hierarchy.

The current online state of a MO corresponds to a set of local neural networks that change and modify, adapting to changing conditions and specific phases of the execution of organism programs. Each network has its own time of fixed existence (TFE), which ensures the dynamic stability of the organism. Just as in the case of the unicellular PS, the MO exists under conditions of dynamic nominal stability, while at the same time fulfilling regular organismic phases and stages and at the same time reacting to stochastic perturbations of the external.

TFE corresponds to the traditional concepts of memory used when considering the functioning of the NS.

Memory of Neural Networks

The time range of the fixed existence of neural networks (the memorization time of their modified state) is very wide from seconds to decades. Short-lived networks perform operational control programs that provide online solution. The most long-lived networks provide, in particular, the strategic age stages of the general organism program.

Currently, various mechanisms for modifying neural networks are being actively studied.

In particular, the effects on the neuron response of the strength and duration of the signal stimulus are studied. For example, the case is considered when a strong and prolonged depolarizing stimulus leads to prolonged impulses. As a result of the action of each pulse, a small amount of Ca^{2+} ions pass through potential-dependent calcium channels into the cell, so that their intracellular concentration gradually rises to a high level. This leads to the discovery of Ca^{2+}-dependent potassium channels, and the permeability of the membrane to potassium increases, which makes depolarization more difficult and increases the intervals between successive pulses.

Thus, with prolonged exposure to a constant stimulus, the strength of the neuron's response gradually decreases. This phenomenon is called adaptation.

Thanks to the adaptation, the neuron, as well as the NS as a whole, is capable of reacting with high sensitivity to a change in the stimulus, even if it occurs against the background of strong constant stimulation.

Due to such adaptability, for example, we do not notice the constant pressure of clothing on our body, but at the same time, we quickly respond to a sudden touch.

Effects that are mediated by receptors and are not associated with channels differ in both duration and delayed expression. This is largely determined by the special role of such effects in the regulation of behavior. They cause a lasting change in the immediate response of the NS to signals coming from outside, and probably form the basis of at least some forms of memory.

For example, the Aplysia mollusk retracts the gill when something touches the siphon.

After repeated touches, the animal habituates and the reaction disappear. In terms of its biological function, habituation is similar to adaptation, but it develops more slowly.

Any sharp stimulus, for example, a strong push or electric shock, removes the effect of habituation and, conversely, increases the sensitivity of the animal, so that it now reacts particularly vigorously to touching the siphon. This sensitization effect persists for many minutes or even hours depending on the strength of the stimulus that caused it, and is a simple form of short-term memory. If the effect of a painful stimulus is repeated for several days, sensitization (i.e., the manifestation of memory) becomes long-term and lasts for several weeks.

Touching the siphon leads to the excitation of a group of sensory neurons. These neurons form excitatory synapses on the motor neurons responsible for gill drawing in. The bases of behavioral phenomena are changes in these synapses. During habituation, the magnitude of the PSP in the mentioned motor neurons decreases upon repeated stimulation by their sensory cells. The opposite effect is observed while the process of the sensitization is at work, we can see the PSP increases.

In either case, a change in the magnitude of the potential is the result of a change in the number of neurotransmitters released from the presynaptic terminals of excited sensory neurons. Thus, the problem is reduced to the question of how the release of the mediator in these synapses is regulated.

In mammals, it is believed that the hippocampus plays a special role in the modification and subsequent memorization of altered neural networks.

In some hippocampal synapses, with multiple repetitive excitations, pronounced functional changes occur. While random unit potentials of action do not leave a noticeable trace in the postsynaptic cell, a short volley of consecutive discharges leads to long-term potentiation, and subsequent single impulses arriving at the presynaptic terminal cause a much greater force in the postsynaptic cell. Depending on the number and intensity of the bursts, the effect lasts for several hours, days or weeks. Potentiation occurs only in activated synapses: synapses on the same cell, remaining at rest, do not change. But if at the same time how one group of synapses receives a volley of consecutive impulses, a single action potential is transmitted through the other synapse on the same cell, then this last synapse also has a long-term potentiation, although a single impulse that came here at another time will not leave a stable trace. This mechanism underlies associative learning.

Other memory mechanisms are discussed.

So, the state of excitation or inhibition in the nerve center can be maintained by impulses wandering in the nerve circuits – lingering on long paths of transmission or returning again to the neuron through closed neuron circuits.

Other researchers suggest that long-term preservation in the nerve cell of a modified state with all the characteristic properties of the stimulus is based on a change in the structure of the proteins making up the cell (possibly also glial cell proteins).

According to the biochemical theory of memory, in the process of memorizing, structural changes occur in ribonucleic acid (RNA) molecules, on which the formation of cell proteins depends. As a result, "altered" proteins begin to be synthesized in the nerve cell with imprints of previous stimuli that constitute the biochemical basis of memory.

EXTERNAL PROGRAMS

EPs along with internal programs make up the PS of the body and solve problems associated with the interaction of the body with the external environment.

EP, most often referred to in traditional literature as behavior, can be divided, for example, into two groups:

- Main, basic
  - Nutrition
  - Breeding (and raising young ones for mammals)
- Additional
  - Sanitary (mainly skin cleaning)
  - Recovery (rest, sleep)
  - Protective (against enemies, adverse weather conditions and other sources of hazards and threats)
  - Arrangement of the site for performing the EP (nests, burrows, marking of the site territory)

MOs carry out all these programs individually or as part of social groups: families (pairs), flocks, etc.

The main makers of EPs are sensory organs, for example, visual, auditory and olfactory, as well as organs of movement and seizure, which are activated by muscle CMs.

The basic tasks of the body, for example, nutrition are solved when performing a complex of external and internal programs. In this case, we can talk about a complex task program implemented by a complex maker consisting of external and internal makers of different levels.

As we have already mentioned, the complete task program of cellular nutrition in a MO includes three local task, composite programs.

This is primarily a program of extraction from the environment of suitable food and its preliminary fragmentation, which is carried out outside the MO, being thus external, and implemented by external organs – makers, which include organs of movement, organs of capture and preliminary fragmentation of a food object, as well as sensory organs.

The other two internal programs bring FFs with the help of makers of the digestive system to the level of components consumed by the cells, i.e., proteins, amino acids, fatty acids, microelements and deliver these nutrient components by the transport makers of the circulatory system from the digestive system to the extracellular space, from which the cells absorb them through their external endocytosis program.

We reviewed the last two internal local NPs of the MO in previous sections.

EXTERNAL NUTRITION PROGRAMS

And now it's time to dwell on the outer part of the NP that solves the problem of extraction and pre-fragmentation of food.

The beginning of NP is carried out by signals from visceroceptors that signal the state of the internal organs, the depletion of the body's energy resources, primarily glucose and fat. A general humoral hunger signal is produced, which triggers, or rather, probably activates, a complex of local task EPs that search for food, capture food material and produce FFs through the reticular formation of the nonspecific NS.

The local task program – the search for food – a food object, which can be defined as auxiliary, preparatory, consists of the following composite programs.

- The first component program – moving to the site of the location of food objects, plant or animal origin
- Next is the searching and finding a specific object
- Finally, the program of approaching the object

The following are the main parts of the external power program:

- Capturing a food object
- Getting FFs

The longest and most energy-intensive local EP is an AP for the search for a food object (SFO).

The complex makers of all the constituent programs included in this local program are external sensory organs, primarily visual, auditory and olfactory; analytical nerve centers and formations of the spinal cord and brain; effector muscle makers of the motor organs of the limbs; proprio and vestibular receptors, fixing states, positions of muscle makers.

The first part of the SFO program begins – the program of moving to the site of the location of food objects by activating the local sensor network corresponding to the memorized landmarks on the way to the nutrition zone.

Sensory signals, for example, visual, received by receptor CMs are further processed sequentially by operational analytic interneurons until a characteristic image of the next reference point is obtained.

If this image corresponds to the pre-memorized one, then, getting to the appropriately configured CN, it launches a set of effector motor organs leading to the next reference point.

It is a possible step-by-step movement of the MO in the zone, i.e., to the site of the location of food objects.

Once in this zone, the MO carries out the following part of the SFO – the search for a specific food object. Carrying out the sensory observation program in this case, the MO scans the situation in the feed zone until the object is in the field of view according to its primary features, for example, in size and contrast, similar to the previously memorized power object.

In this case, the sensory information of such an object is processed, analyzed by the operating neurons further to the level corresponding to the pre-stored image of the food object.

This information about the detected object falls on the CN that is configured on the food object, and if it coincides with this setting, it triggers the neuron. When triggered, the neuron launches the effectors of the next stage – of approaching the detected food object.

This program is performed in the same way as chemotaxis of unicellular, when the motor part of the program works out the movement of the organism in the direction of the gradient of the receptor, sensory signal coming from the target of the movement, in this case the food object. Such a movement can be represented as a sequence of program steps, each of which in turn includes the following phases:

- Sensory fixation of the current position of the object
- Comparing it with the memorized position of the previous step
- Elaboration of motion parameters of a step (direction and speed) based on the results of comparison
- Memorizing the current position for comparison in a subsequent step
- Implementation of the current motor step

After which these phases are repeated in the next step of the program.

In the case of a fast and unevenly moving object, for example, a predator's victim, the rendezvous program is significantly more complex and includes the steps of covertly observing the victim, stalking it and bringing it directly to contact.

The main parts of the external NP – the capture of a food object and the receipt of FFs are performed using the gripping extremities of the maxillary and motor makers of MO. These programs are performed with the participation of both exteroreceptors and interoreceptors. As in all other EPs of different structural levels, processing and analyzing programs of interneurons actively work in them.

Program Space (PS)

PS is a part of the external environment in which organism, animal exists, functions, lives, i.e., executes its EPs.

PS includes program execution sites and those structures that the animal itself creates, such as nests and holes, transport routes between sites; program facilities: food sources, sexual partners, single species neighbors, hazards, threats.

An animal fixes the borders of the own PS with the help of special marks that are secreted from the corresponding glands.

Mandatory attributes of the PS are signs, landmarks, symbols, by which the animal finds the site of the current program. At the same time, in the animal's NS, there seems to be a connection between the geographic symbol, the landmark of the program execution site and the launch signal of the same program.

The combination of such symbol-signal connections creates a standard, regular representation of the PS for the animal.

PS can change. There are two types of changes. The first type is predictable or expected changes. They are most often associated and caused by the change of natural cycles, mainly diurnal and seasonal. The second type is stochastic unpredictable changes due to natural disasters or the invasion of competing or predatory animals.

Stochastic unfavorable changes are eliminated and corrected by animals through the implementation of the corresponding mainly building programs according to signals about violations of the standard, regular PS. A vivid example of this is the reaction of ants to a destroyed anthill.

Expected, standard changes set the internal rhythms of the animal, which determine the pace and sequence of execution of both internal and EPs. Especially important for the program schedule of the animal are circadian rhythms.

In some species, there are two geographically separated PSs for various external feeding and breeding programs, for example, PS of birds, spawning fish and some species of insects (butterflies).

In the case of a permanent presence in one PS of several representatives of one species, the collective, social use of this PS is carried out. Thus, there arises the social behavior of animals, which determines the mutual order and sequence of EPs for different members of society. The features of social behavior are the allocation of PS sub territories for individual members of society, priorities in food, etc.

Structure of External Programs

The task EP consists of two stages: preparatory, auxiliary (observation) and the main (the task program itself). The most important role is played by observation in the programs of nutrition, reproduction and arrangement of the program site.

The local tasks of the first stage are to establish, clarify the position, localization of the main program execution site (usually within the program space) and monitor the objects of this main program and their environment until the start of the main program.

The observation program includes sensory programs, usually visual, auditory and olfactory, and, transforming, processing and analyzing of sensory information to a level that can initiate the action potential of the CN that launches the main program.

The monitoring program occurs frame by frame. The current observable situation, for example, the visual (online frame) perceived by the primary photoreceptors, is processed by the analytical interneurons to the level of the image capable of initiating the CN triggering the initial stages of the main program.

If such a launch does not occur, then the situation is "not ripe" to start the main program and the observation continues by registering the next frame until fixing the "ripe" situation and then starting the main program and terminating the monitoring program.

At the same time, each situational frame is remembered in an operational and not in a long-term manner. Each current frame is compared with the previous memorized one and a corrective control action is generated on the neuron controlling, for example, by the movement and position of the eyes.

During the execution of the information observation AP, two tasks, tactical and strategic, are performed.

The tactical task is to capture the sensory situation to trigger the CN that launches the main program directly in the current implementation of this program.

The implementation of a strategic task leads to the refinement, correction of both landmarks in order to find a site, and the position and behavior of objects necessary to increase the effectiveness of future implementations of the task program.

Task EPs usually consist of several, often many, local component programs that form a hierarchical series-parallel structure. For example, the program for finding and capturing a food object consists of many composite programs and components of the movement programs of the entire MO as a whole, and its individual parts, the orientation of both the MO and its constituent organs of motion and capture, etc.

Sensor, Analytical and Effector Blocks of External Programs

When implementing all the task-oriented EPs and their components, the following set of program actions is carried out:

- Receptor makers, recording information about the external environment, primarily about program objects and the status of their execution sites
- Analytical neural makers processing receptor, sensory information to the level of perception by CNs
- Effector motor makers that launch and implement motor parts of programs

We can see that there are three blocks of EPs and their makers: sensory, analytical and effector.

At the same time, the analytical unit, being an intermediate one, on the one hand, forms common structures with both the sensor unit and the effector motor part of the motor unit.

Work Scheme and Interaction of Blocks

Let us consider schematically the work and interaction of the sensory, analytical and effector blocks, based on information received from the visual sensor.

In the task functional visual maker (TFVM), in addition to the network of neurons of the retina, primary and secondary visual cortex, we also include the opto-mechanical part forming the visual organ – the eye.

The eye OM performs particular programs of focusing the detecting image on the receptors, regulating the intensity of the image and adjusting its position relative to the retina containing these receptors.

In general, the TFVM performs the task of forming the current visual image of the situation on the site of the EP and comparing it with the memory of the successful previous implementations of this program.

Let us consider briefly an example of the scenario of the program executed by the TFVM at the observation stage.

The recorded image enters the eye through the pupil, a hole in the iris, the diameter of which is regulated by two muscular antagonist makers: the sphincter narrowing the pupil and the dilator expanding it.

These makers are controlled by an integral signal from the retina receptors. Thus, the light intensity is set on the retina, which is comfortable for recording the image with the maximum spatial resolution.

Then the light passes through the lens crystalline lens, the focal length of which with the help of adjacent muscle makers is set depending on the distance to the object in question by signals from the TFVM analytic neurons that react to the boundaries of the object and its parts.

Programs performed by the iris and lens as effector makers are implemented with the participation of receptor and neural CMs.

Next, the recording image falls on the first receptor layer of the eye retina, where it is converted into a set of nerve impulses at the receptor synapse, as described in the previous section.

The first stage of image processing is the selection of basic information elements from it: dark and light spots, and contrast strips located at different angles. It's performed by neural CMs of the retina and primary visual cortex. This type of processing is often called differential analysis in general considerations of human mental operations.

The second stage is the compilation of a single information visual image from the selected image elements, which is most often the outline of objects, carried out to a large extent by neural cellular makers of the secondary visual cortex. In the same place, the current information image is compared with the memorized ones in previous program implementations. This second stage corresponds to the term "synthesis of information".

And when comparing the current image with the previously remembered in different situations and in accordance with these situations, separated into different groups of images, the third most important action of image processing and the human mental operation corresponding to it – image recognition is carried out.

The three most important program stages of image processing that we have listed are as follows: analysis, synthesis and recognition are inherent in all MOs. Together with the recognition of their essential analogy with the basic human thought processes, we receive another argument in favor of the unity of the information organization of PSs at all structural levels of organisms.

It can be assumed that the recognition option is the triggering of one of the CNs that launch their local task programs when the synapses of these neurons receive signs of the image to which they were configured by previous program implementations. At the same time, signals from the signs of the current image by divergence arrive simultaneously on several CNs tuned to different remembered images and launching various subsequent programs. Recognition of the current image with the launch of the subsequent program is implemented on the CN, the preset of which corresponds to this image. In this case, we can say that the result of the image-recognition program of the current situation is the launch of the next local program specific to this situation.

When performing observation, examples of such programs can be the following.

Continued observation of the detected "necessary" objects, but located too heap, inconvenient for subsequent pursuit and capture of one of the objects during the execution of the hunting program.

There is activation and initiation of the follow-up local programs of an object pursuit and seizure in the event of a "ripe" situation, when the desired object of hunting is localized and within reach.

Finally, in the absence of a hunting object in the observation zone, the program of searching for objects, first by scanning the previous observation zone and then the expanded zone, is launched.

In addition to the sensor and analytical blocks, all three examples of launched programs also involve effector and motor ones.

In the first case, the periodic fixation of the situation is accompanied by both corrective movements of the eyes, head and torso of the observing MO, directed by careful slow movements of the MO toward the object.

In the second example, in the pursuit program, the movement of the MO in the direction of the object is realized at speeds close to the limit.

In the program of scanning the observation zone, cyclic motions of the MO in this zone are performed, as well as the corresponding rotations of the necks and eye rotations.

In all these cases, the main active makers of motor programs are tissue muscle makers, consisting of cellular makers – muscle fibers.

The long, thin muscle fibers that make up the skeletal muscle are gigantic cells that are formed at the fusion of many individual cells during ontogenesis. They are multi-core, and the nuclei are located directly under the plasma membrane. The main part of the cytoplasm consists of myofibrils – cylindrical elements with a thickness of 1–2 microns, which often stretch from one end of the cell to the other. As noted in the previous sections, the reduction of muscle fibers and, accordingly, the entire muscle consisting of them, occurs due to the interaction of actin and myosin filaments, activated during the hydrolysis of ATP energy molecules.

Muscle fiber contraction occurs when an excitation wave arrives in the form of nerve impulses along the axons of motor neurons to the neuromuscular synapses.

Muscle stimulation leads to the release of calcium from the sarcoplasmic reticulum and its connection to the regulatory protein that allows the contractile proteins actin and myosin to interact.

Any muscle fiber is innervated by only one motor neuron. At the same time, one motor neuron innervates a group of muscle fibers, with which it forms a motor unit. It is the smallest functional unit of the motor system. In different muscles, motor units differ greatly in the number of muscle fibers. So, for example, in the oculomotor muscles there are about a dozen muscle fibers per one neuron, in the motor units of the small muscles there are about a hundred and in the large muscles of the limbs and body there are more than one thousand per motor neuron. Small units are characteristic of the muscles with which precise, well-coordinated movements are performed, while the larger ones prevail in massive muscles, such as, for example, the back muscles.

The number of activated motor units can regulate the strength of muscle contractions. Such mechanism of regulation is called recruitment.

Skeletal muscles attach to the bones and, in their contraction, change the position of a joint. Muscles that produce movement in one direction are called agonists or synergists, while those providing movement in the opposite direction are called antagonists. For example, all the muscles, the reduction of which leads to flexion of the limb in the joint, will be synergists, and the muscles that provide extension are antagonists for them.

The typical task movement maker (TMM), which forms the so-called musculoskeletal system of the MO, in addition to the active muscle component maker, also includes "passive" components that transfer energy of muscle contractions to the final TMM elements that directly interact with the object of the motor programs. Examples of such finite elements can be the ends of the limbs that serve to move, the feet and toes (paws) or the jaws of the MO.

The main passive components of TMM, which convert muscle contractions into the movement of finite elements, include a system of bones interconnected, as a rule, with movable hinges – joints and tendons that have high elasticity and elongation and transmit muscle contractions on the skeleton bones adjacent to the muscle.

Bone is a matrix that consists mainly of a composite material that includes inorganic mineral calcium phosphate, which gives the main bone stiffness and collagen – elastic protein. Bone is in a constant process of development and destruction. To do this, it contains osteoblasts – bone-forming CMs, and osteoclasts – cells that destroy bone in order to prevent it from excessive thickening.

The composition of the joints and tendons includes collagen formed by special cellular makers – fibroblasts.

Muscle neurons of the spinal cord and cells of the motor nuclei of certain cranial nerves control the activity of the muscles and, therefore, the motor makers. They can be activated by sensory neurons (in the implementation of monosynaptic reflexes), but in most cases, the nearest excitatory and inhibitory interneurons determine the activity of motoneurons.

The functional association of motoneurons with neighboring interneurons, designed to control a specific part of the body, is a lower motor system or a local motor apparatus. Such structures control different parts of the body: hand, foot, eye. Different muscles are involved in each individual action, some of which contract, while others relax at the same time, resulting in, for example, flexion of the arm or extension of the leg.

In any lower motor system, as in a card index, there are programs of all possible movements of the controlled part of the body; therefore, the task of the command motor centers is to select the desired program from this card index. In the simplest case, when the movement is performed according to the programmed mechanism of the spinal reflex, this choice is made by the local motor apparatus itself, depending on the nature of the sensory information. It can, for example, interrupt the bending of the fingers that take an object if this object turns out to be very hot. The choice of a specific motor program, as a rule, is determined by the most significant sensory information and most often it consists in the preference for the most effective actions.

Each motor center, at whatever level it is, receives its own quota of sensory information. The local motor apparatus neurons use the sensory flow from the receptors of the muscles, tendons and joints, from the superficial and deep receptors of the skin and from the interoreceptors of the internal organs. The trunk motor centers, along with this information, use signals from the vestibular, visual and auditory receptors in their activities. The motor cortex receives the sum of the necessary information from the sensory cortex, and in addition, it depends on the associative regions that have already integrated all types of sensory information. The continuous flow of sensory information at all levels of the organization of motor systems in a timely manner provides each motor structure with operational feedback, i.e., information about how a particular movement is performed, whether or not the intended goal is achieved and according to these data, the movements performed are constantly adjusted.

The hierarchy of motor centers is manifested in the fact that the higher ones can cancel the lower commands or instruct them to execute their own command. For example, stem motor centers are able to subordinate the activity of the lower motor systems of the spinal cord, but they themselves are forced to obey the motor areas of the cortex.

Possible Principles of Formation and Inheritance of External Programs

Depending on the hierarchical level of the program, it can solve the task of finally bringing the motor maker to the object, which is carried out with the constant exchange of information between the sensors and the motor maker with the involvement of a limited analytical unit. As a rule, a simple neural network of such a tactical program as well as the components of its composite maker change little over the lifetime of the MO and are formed at the stage of ontogenesis in accordance with the gene information.

In other strategic cases, the analytical unit makes a decision about the readiness of the situation in the program space to issue a trigger signal to the motor neuron after processing a large amount of sensory information of often different modalities. In this case, the motor maker of such a behavioral program of a high hierarchical level is usually a combination of the motor makers of the local components of the programs. Accordingly, the composite maker of the analytical unit performs a multi-level SPs that process and analyze information of various modal sensors and collect the required simple components of the motor programs. The neural network of such a strategic task program is much broader and more developed than in the first tactical case considered.

Since the program space with objects of EPs, for example, food or sexual partners, is subject to constant changes, it requires constant monitoring of these changes by sensory OMs, analysis of these changes and variation, modification, correction of effector, motor programs in accordance with the new sensory information.

The sensors record a qualitative change in the situation with objects and conditions around, primarily within the program area. In accordance with these changes and the results of their analysis, the complex of effector motor programs is modified. In other words, for a new sensory situation, a new modified adequate motor complex is created. In general, this implies modification, updating of the corresponding local neural network.

In fact, this adaptation of the EP to new conditions largely corresponds to the learning process considered in the works on the behavior of organisms.

It should be noted that the participation of the hippocampus both in the formation of emotional reactions and in the processes of long-term memorization suggests, for example, that the emotionally marked successful result of the modified program contributes to memorizing the configuration corresponding to its updated neural network.

When learning, the memory mechanisms discussed in the previous section are used, associated with changes in the effectiveness of synaptic contacts depending on the frequency of the signals passing through them. Neural networks that are rarely used weaken and disappear, while frequently used ones become more efficient and accessible to use.

At the same time, the input neural (dendritic) network of the CN with active synapses formed in previous successful program implementations (or innate) corresponds to the program space situation with the food objects and their environment adequate (matured) for launching the motor program block of this task programs.

The axon of a given CN forms the action signals specifically for those motor effector makers that are involved in this program.

So, the program reproduces the successfully tested or innate response of an adequate motor activity to a standard, typical sensory situation.

In behavioral EPs of a higher hierarchical level, a CN starts not specific motor makers, but the above-described local PSs from sensory, analytical and motor blocks.

PROGRAM AND INFORMATION STRUCTURES OF MULTICELLULAR ORGANISMS

MO is a combination of cells reproduced anew in the life of each organism from two initial germ cells. MO is created for the successful evolution of precisely these "immortal" germ cells, their PS. PS in the initial starting form is encoded in the genetic material, compartments and cytosol composition of these cells. In accordance with the gene instructions and the initial composition of the makers, specific cell divisions, their differentiation and transport, which lead to the structuring of MOs, occur in the organism.

Thus, MO is an evolutionary representative of the expanded PS of the reproductive cell and defends its viability in the fight against stochastics of the outside world, biological enemies and competitors.

The successful conduct of this struggle, as well as the implementation of reproduction of both sex cells and their representative, MO, is the main evolutionary goals of the developed PS of a mature, but not old MO, which we are considering.

Thus, outside the scope of this book remains the stage of ontogenesis – the formation and development of MO.

When listing all the main task programs of a MO, it is necessary to take into account the programs implemented directly in the germ cells and we will start with them.

Germ cells ensure continuity of generations. In many animals, such as insects and vertebrates, there is a clear and early separation of the germ cell line from the somatic cell line.

In organisms with a fairly early isolation of germ cells, the latter do not arise in the gonad itself. Their predecessors, primary germ cells (PGCs), migrate to the developing gonad. The first stage of gametogenesis, therefore, includes the colonization by the PGCs of the genital ridge of the emerging gonad.

In mammals, PGCs migrate along a path very similar to that seen in tailless amphibians.

Upon reaching the gonad, PGCs are divided and form spermatogonia of type A1. These cells are characterized by smaller sizes and oval nucleus, in which chromatin is associated with the nuclear membrane.

The next stage of sperm maturation is spermatogenesis – sperm cell differentiation. In order for fertilization to occur, the sperm must meet the egg and unite with it; therefore, in the process of spermatogenesis, the sperm undergoes differentiation, allowing it to acquire the ability to move and interact with the egg.

If a gamete formed as a result of spermatogenesis is essentially a mobile nucleus, then a gamete formed during oogenesis has all the factors necessary for initiating and maintaining metabolism and development. Consequently, in addition to the haploid nucleus, a pool of cytoplasmic enzymes, matrices, organelles and metabolic substrates occurs in ontogenesis. Thus, if sperm differentiation is aimed at motility, then the development of an oocyte leads to an extremely complex organization of the cytoplasm. Unlike sperm, the differentiation of which ends after the division of meiosis, the growth of the oocyte occurs primarily during the extended period of the prophase of meiosis.

Gametes of a single species differ sharply in size and mobility into two types: small mobile male gametes – spermatozoa – and large stationary female gametes – ova.

The program scheme of a MO, including the hierarchy of makers, was presented at the beginning of the chapter. Here we once again briefly list the task programs MO.

The basic task programs performed by somatic cells individually, or as part of the organs and systems of MO, include, first, programs of nutrition and respiration, reproductive program, protective IP, as well as sanitary excretory programs of the kidneys and colon. The most important social programs for the exchange of material resources and information signals between the makers of all levels are carried out by the circulatory and NSs of the MO.

In previous sections of this chapter, we specifically considered these programs, with the exception of reproductive and sanitation programs, omitted for reasons of adherence to the principle of "necessary sufficiency" of factual material.

Next, we will try to show how the cellular information system, discussed in the last section of Chapter 1, works, modified and developed in a multicellular social organism.

It seems to us that the main differences between the MO PS and the single-cell PS are its social structure and the development of EPs.

It is on the manifestations of these features in the information system of the MO that we will end our consideration, based on the material of the previous sections of this chapter.

All somatic cells of MOs, as a rule, go through several stages of differentiation as tasks change for their elementary social makers and for these cells as a whole, acting as CMs.

Here we note that, in essence, the differentiation of MO cells is very close to the phenomenon of metamorphosis, which was examined in detail for *Naegleria gruberi*. For most of its life cycle, *N. gruberi* is a typical amoeba. However, when the suspension of bacteria is diluted, each individual *N. gruberi* rapidly acquires a streamlined shape, and two long flagella develop at its front end. The transformation of the amoeba into a flagellate is carried out in 1 hour. During this time, both the kinetic apparatus, which would serve as basal bodies for flagella, and flagella should appear in the amoeba. Flagella are built of many proteins, among which tubulin predominates. Tubulin molecules are organized in the form of microtubules that are combined into a structure capable of supporting the movement of flagella. At the stage of amoeba tubulin for flagella, Naegleria is absent. It is formed de novo after the start of DNA transcription and is a great example of development control at the level of transcription. The Naegleria core responds to environmental changes by synthesizing mRNA for tubulin flagella.

Thus, the task program of metamorphosis includes compound programs that significantly modify the amoebic SPs. As we noted at the end of Chapter 1, such notable rearrangements of the PS are controlled and monitored by the cellular signaling network with the main role of intelligent makers – deciders and, above all, their "heads" of the p53 protein. In this case, the main task of such a signaling network is to enable the expression of chemotaxis genes.

The program of metamorphosis is started by an external signal of dilution of amebic suspension.

Organizing the metamorphosis program is organizationally very reminiscent of the implementation of the cell division phase in response to an external proliferative signal, also discussed in Chapter 1.

Based on the foregoing, it can be assumed that the step-by-step differentiation programs of MO cells are implemented according to similar schemes under the control of signaling cellular networks and, above all, their makers, the deciders.

Let us further consider some of the main features of the development of an information network in a MO, related to the social structure and the large volume and importance of EPs of MO.

As we have repeatedly noted in this chapter, social makers of such an organism form a multistage hierarchical structure that includes molecular "elementary" makers (EM), CMs, TMs, OMs, SMs and, finally, a complex fusion of makers (CFM) to perform a complete task program, for example, a NP. TM, OM, SM and CFM consist of EM and CM and therefore we will call them composite.

Despite the fact that in all the makers of the highest hierarchical level, the "real" work on the execution of programs is ultimately performed by EMs, the above structure is not formal. It is significant that in many cases the EM and the CM execute programs consisting of makers of a higher order synchronously with each other and in coordination with those included in other composite makers.

At the same time, structural makers of CM and OM perform important functions for makers of the lower hierarchy: CM for EM and OM for CM. They include sites for performing basic programs of junior, lower makers, and also contain auxiliary makers that, first, promote reproduction, proliferation and/or maturation, formation of basic makers.

The reader was acquainted with the real work of makers at all levels when considering examples of performing specific task programs of the MO.

Its NS is also arranged hierarchically to control the multistage structure of makers in the MO.

Inheriting the single-cell chain management structure, MO informational makers also perform three types of information programs: initial – receptor, final – effector and intermediate – summing, processing and decisive.

The reader was acquainted with the real work of makers at all levels when considering examples of performing specific task programs of the MO.

To control the multistage structure of makers in the MO, its NS is also arranged hierarchically.

In this case, the standard task management program is collected, which is collected from these local programs.

First, receptor, sensory information makers collect the primary signaling situation corresponding to this task program.

Then, this primary signaling information is processed and summed up by intelligent makers until the information situation is obtained, which is necessary and sufficient for making a decision on launching the final effector local program.

And, finally, an effector neural maker runs this program.

A group of information makers connected into network structures and forming composite task makers performs the overwhelming part of local task programs.

And, finally, an effector neural maker runs this program.

In accordance with the modern view of the functional structure of the NS, it consists of local neural networks formed in its peripheral and central parts, in accordance with the tasks that they solve – from the simplest reflex to complex behavioral. At the same time, networks performing complex behavioral programs consist of many simple and more complex sensory, effector and processing, analytical networks of various levels of complexity and hierarchy.

Network structures can be divided into virtual, temporary, formed at the time of solving this problem, mainly when performing EPs, and anatomical, constantly existing in the body of neuron clusters. These latter include, in particular, the nerve centers, ganglia and nerve nodes, the spinal cord and the brain, each with its own network structure, that are listed according to their participation in increasingly complex behavioral programs.

In this case, anatomical innate networks perform two main functions. First, they directly autonomously solve many of the so-called reflex tasks of the lower level, for example, the ganglia in the digestive tract control its peristaltic and absorption programs. And second, they provide the initial basic set of neural clusters and the main connections inside and between them for the subsequent creation of actual virtual network configurations.

Depending on the specific task, the degree of involvement of different levels of network neural structures varies.

For example, an arbitrary change in breathing is carried out by a person with the participation of the centers of the cerebral cortex, a special center in the midbrain, the respiratory center, the medulla oblongata and the centers of the spinal cord that innervate the respiratory muscles.

Two mechanisms are used to form network structures.

The first of these, convergence, consists in integrating and processing by a single neuron, as a rule, by an intelligent neural maker (n-decider), of a variety of input signals arriving at its dendrites, for example, from individual receptors or a preprocessed information from various sensors of visual, auditory, olfactory and other. On the axon of such an n-decider, the resulting signal arises, which is sent further to intelligent neurons or directly to effector makers if the configuration of the input signals matches its preconfiguration.

In this case, a second network mechanism, the divergence, may be involved when the signal is sent simultaneously to the inputs of various "consumers" along a set of axon branches (collaterals).

For example, large human motor neuron innervates dozens of muscle fibers in the outer eye muscles and thousands in the muscles of the limbs.

We see that the formation and functioning of neural networks of MOs are carried out on the same basic principles on which intracellular network information structures operate, and a highly intelligent n-decider with multiple inputs and outputs is very similar to p53.

Apparently, the network information system of multicellular, especially among the highest representatives of mammals, has received the greatest development due to the growing diversity and evolutionary significance of EPs that are very limited in single-cell ones.

The totality of individual receptor makers was transformed into sensory OMs: visual, auditory, olfactory and others, issuing multi-bit signal patterns about the situation on execution sites of EPs.

Accordingly, intelligent local analytical networks of n-deciders have developed to transform and analyze signal sensory information and control effector muscle makers.

Let us briefly recall a schematic example of the information network's operation from the visual sensor to muscle effectors.

The recorded image falls on the first receptor layer of the eye retina, where it is transformed into a set of nerve impulses at the synapses of the receptors.

The first stage of image processing is the extraction of basic information elements from it: dark and light spots, and contrast strips located at different angles, performed by neural CMs of the retina and primary visual cortex.

The second stage is the compilation of a single information visual image from the selected image elements, which is most often the outline of objects, carried out to a large extent by neural cellular makers of the secondary visual cortex.

In the same place, the third stage of composite analyzer visual maker's work is apparently carried out – a comparison of the current information image with the ones stored in previous program implementations.

A recognition option may be the triggering of one of the intelligent neurons that launch their local task programs when signs of the image that they were configured in previous program implementations arrive at its synapses.

It can be assumed that such a decision is made to launch an adequate effector program in response to a specific situation observed by sensors on the execution site of the EP.

Signals from the signs of the current image by divergence simultaneously arrive at several intelligent neurons that launch various subsequent programs and tuned in to the various memorized images. The implementation of recognition of the current image with the launch of the subsequent program occurs on the neuron, the setting of which corresponds to this image. In this case, we can say that the result of the current situation image-recognition program is the launch of the following local program, which is specific and adequate for the given situation.

The recognition result and possibly additional analyzer operations should be a composite local task program executed by effector muscle makers. This program, adequate to the observed situation, is controlled by a section of the motor neural network, the so-called local motor apparatus, that selects the anatomical reflex innate motor networks of the lower level needed to solve the current problem.

Returning to the general consideration of the MO information system, it can be said that each control task has its own program executed by a composite network neural maker. The current online state of a MO corresponds to a set of local neural networks that change and are modified, adapting to changing conditions and specific phases of the execution of organism programs. In addition, each network has its own TFE, which ensures the dynamic stability of the organism. Just as in the case of the unicellular PS, the MO exists under conditions of dynamic nominal stability, while at the same time fulfilling regular organismic phases and stages and at the same time reacting to stochastic perturbations of the external environment.

The time of the fixed existence of the neural network (TFE) corresponds to the traditional concepts of memory used when considering the functioning of the NS.

The time range of the fixed existence of neural networks or using the generally accepted terms of the memorization time of their modified state is very wide from seconds to decades. Short-lived networks perform operational control programs that provide online solution. The most long-lived networks provide, in particular, the strategic age stages of the general organism program.

Currently, various mechanisms for modifying neural networks are being actively investigated, as a consequence of a change in the response of a neuron to a signal.

For example, it is known that with prolonged exposure to a signal, the intensity of the response of a neuron gradually decreases. This phenomenon, called adaptation, is caused by the action of potential-dependent calcium channels and Ca^{2+}-dependent potassium channels.

When exposed to a sharp stimulus, for example, a strong push or electric shock, the effect of habituation, adaptation decreases and the sensitivity of the animal increases. This effect of sensitization persists for many minutes or even hours depending on the strength of the stimulus that caused it and represents one of the forms of short-term memory.

In both cases, the change in the response of the neuron is a consequence of varying the intensity of the release of the mediator in its synapses.

In mammals, it is believed that the hippocampus plays a special role in the modification and subsequent memorization of altered neural networks.

In some hippocampal synapses, with repetitive excitation occur pronounced functional changes. While random unit potentials of action do not leave a noticeable trace in the postsynaptic cell, a short volley of consecutive discharges leads to long-term potentiation, and subsequent single impulses arriving at the presynaptic terminal cause a much greater force in the postsynaptic cell. Depending on the number and intensity of the bursts, the effect lasts for several hours, days or weeks. Potentiation occurs only in activated synapses: synapses on the same cell, remaining at rest, do not change. But if at the same time how one group of synapses receives a volley of consecutive impulses, a single action potential is transmitted through the other synapse on the same cell, then this last synapse also has a long-term potentiation, although a single impulse that came here at another time will not leave a stable trace. It might be that this mechanism underlies the associative learning.

Other memory mechanisms are discussed.

So, the state of excitation or inhibition in the nerve center can be maintained by impulses wandering in the nerve circuits – lingering on long paths of transmission or returning again to the neuron through closed neuron circuits.

Other researchers suggest that long-term preservation in the nerve cell of a modified state with all the characteristic properties of the stimulus is based on a change in the structure of the proteins making up the cell (possibly also glial cell proteins).

The program space of executing EPs with objects in it, for example, food or sexual partners, is subject to constant changes. Therefore, constant monitoring of these changes by sensory OMs, analysis of these changes and correction of motor programs in accordance with the new sensory information are required.

In fact, this adaptation of the EP to new conditions largely corresponds to the learning process considered in the literature on the behavior of organisms.

The initial stage of training is reduced to the method of trial and error.

It should be noted that the participation of the hippocampus both in the formation of emotional reactions and in the processes of long-term memorization suggests, for example, that the emotionally noted successful result of the modified program contributes to memorizing the configuration of the corresponding neural network.

When learning, the memory mechanisms discussed in the previous section are used, associated with changes in the effectiveness of synaptic contacts depending on the frequency of the signals passing through them. Neural networks that are rarely used weaken and disappear, while frequently used ones become more efficient and accessible to use. In general, the well-known saying is realized: “repetition is the mother of learning”.

3 Man

LEAP FORWARD IN EVOLUTION – EMERGENCE OF HUMAN VOLITIONAL PROGRAMS

Human is the next evolutionary leap behind multicellular organisms (MOs).

The same method of the program approach and, specifically, a separate consideration of its internal and external program subsystems will help us to understand the essence of the qualitative leap that occurred in evolution when a human appeared.

If we look at the internal program subsystem of a human, then we will not notice any significant leaps, and even more qualitative ones, in comparison with highly developed mammals. We can clearly see the same structure and functioning of all systems and organs nutrition, respiration, blood circulation, reproduction and all the rest. The management of internal programs using the humoral and nervous systems is organized and functioning in a similar way.

Moreover, a human has a reflex-intuitive section of external programs, which also has not changed and functions similarly to, say, humanoid primates.

So what is the nature of the leap and how significant is it?

With just a quick glance at the results and methods of external program activities of a person, including the prehistoric one, the answer becomes obvious.

Human himself individually constitutes or at least can compile external programs. In other words, he can set a program task, set a result, choose and, if necessary, make makers and objects for it, ensure its implementation with energy and sites.

At the same time, he uses not only and not so much the program elements developed by the previous evolution, but those that he himself or his "social colleagues" got in carrying out the previous independent (not evolutionarily constructed) programs.

All other MOs, including the most advanced mammals, carry out their external programs within the framework of congenital, evolved capabilities of the previous evolution, determined by specific blocks of sensory, analytical and effector primarily muscle makers, as was shown in Chapter 2. Of course, the MO can modify and correct complex external programs, adapting them to the real external situation, but only using innate reflex base organs and their formations.

The emergence of the established "Homo Programming" can be correlated with the era of Cro-Magnon man, when independent human activity was already fully manifested in all external programs, from hunting to building a home.

A man passed the evolutionary period from anthropoid ape in a few million years, bypassing the stages from Australopithecus to Cro-Magnon.

The question is – what qualitative evolutionary changes appeared in the human body for this period? The modern anthropology gives a clear answer to this question. This is primarily a change in the structure of the brain, which led, in particular, to the appearance of the Wernicke and Broca fields responsible for the perception and processing of the language. Both these structures are already quite clearly expressed on the endocranial cast of fossil skulls of genus Homo representatives.

This aspect of human evolution is significant to us.

Traditionally, language is viewed primarily as a means of communication between people.

In the program approach, its main role is to fix the program information: the scenarios and the above-mentioned program components that have already become standard for us.

For this purpose, in the human language system, informational character marks of the word are used, which designate the information pattern of the image – sensory or effector, similar to the situational images, which we first considered for the information systems (ISs) of the cell and in the expanded use of multicellular.

It is easy to see that in all languages, without exception, basic words denote such program components as makers and objects (nouns), actions of makers on objects (verbs), site – the place of such action (nouns, adverb), as well as signs and characteristics of these components from which their image is assembled (adjectives, numerals, adverb). For example, color, size, hardness of makers and objects, speed and intensity of action, geographical location, site convenience, etc. Messages made up of words correspond to a program scenario.

Thus, by speaking to himself or out loud, the corresponding words and the messages made up of them, a person at any moment of time at will, can virtually reproduce any previously memorized program.

Then he can correct and modify the old program and create a new one from the known stored components.

By reproducing virtually different versions of programs, a person can compare their efficiency, energy consumption, etc., and choose the one appropriate to the situation and the task.

Let us call human programs controlled, modified and created by him – volitional programs (VPs), keeping in mind, above all, their arbitrary, creative nature, in contrast to the overwhelmingly innate, reflex in all other organisms.

Here it should be noted that with the advent of Homo Programming, collective VPs performed by groups of people organized by thematic (hunting, tool making, construction) or family relationships (family members, clan) acquired a large and soon overwhelming role. Such early grouping in the implementation of major programs has led to a sharp increase in the effectiveness of their implementation and accelerated their improvement.

At the same time, individual, personal nutrition programs, maintaining homeostasis, sanitation and hygiene, breeding and caring for offspring, etc., have been preserved for the entire existence of mankind.

Further in this chapter we will note the social nature of human communities, as a direct consequence of the emergence of VPs.

Let us consider the examples of VPs, both individual and collective, we will show some of the principles of their interaction.

We will look at the history of mankind from the Cro-Magnon period to the present day as a continuation of the evolutionary process for the system of VPs.

Let us try to formulate the structure and principles of the volitional programming system and briefly examine its main elements.

Finally, at the end of the chapter, we trace the inheritance and development of ISs of unicellular and MOs in VPs.

In separate sections, we will very briefly discuss the issues of scientific methodology and picture of the world.

SOCIAL STRUCTURE OF HUMAN COMMUNITIES – PERSONAL AND COLLECTIVE PROGRAMS

Autonomous System of Volitional Programs (ASVPs)

As well as for all MOs, human external programs, in our consideration VPs, provide, first, the implementation of internal programs, among which one should single out a set of programs for maintaining homeostasis and specifically the most important nutrition program of the body.

Thus, nutrition and programs that provide protection from the external environment, primarily construction programs, can be called the main basic VPs, which correspond to the most important external programs of all MOs.

In the era of the Upper Paleolithic from 40 to 10 thousand years ago, there was one of the first representatives of Homo Programming – the so-called Cro-Magnon human. His brain had already developed frontal, parietal and temporal lobes that are responsible for specific human characteristics of behavior and psyche, or, in our terminology, for volitional human programming.

It was during this period that the main volitional external programs of human behavior were created, replacing to a large extent the genetic external programs of other multicellular ones.

These are food programs, of which, first, it is necessary to single out the hunt with the use of new tools – makers: traps, darts, knives, harpoons and then a bow and arrow.

Next is the program of habitat arrangements, in particular the creation of sites with capital structures.

Among the programs aimed at maintaining homeostasis, we will outline the manufacture of clothes and shoes.

It is significant that these programs for the most part were compiling and executing collectively, which led to the development of the social structure of community – the tribal one.

In the Cro-Magnon community, which was a socio-geographical localized human community, its own autonomous system of volitional programs (ASVPs) developed and carried out. Contacts with other neighboring communities were not necessary and, depending on their nature, could have a positive or negative impact on the community.

It can be said that ASVP is the essence of an independent, self-sufficient human community capable of independent evolutionary development.

Thus, human history can be represented as the evolution of the ASVP from the Cro-Magnon community to the modern developed state.

Human Social Paradigm – The Relationship between Personal and Collective Programs

The display of sociality in ASVP is essentially similar to that observed in a MO.

A person takes away a part of his resources from his personal programs and spends them on participation in collective ASVP programs, receiving in return social benefits from the possible use of diverse collective results. Naturally, the collective results of the ASVP are fundamentally higher in terms of the level and nomenclature of those that an individual could achieve, performing only his own personal programs.

Thus, the evolutionary prospects of social structures of ASVP are beyond doubt.

A comment should be made here regarding the broader interpretation of the term "sociality" used in this book, compared to what is meant, for example, by the privileges provided by a modern ASVP, by the state to its deprived citizens.

So, personal programs are compiled, modified and executed by each person individually and in turn are subdivided into internal individual, as a rule, composite programs and external programs that are components in collective programs.

Collective programs are composite and consist of external personal programs.

In personal individual VPs, it is possible to identify basic ones that include food, reproduction, programs providing homeostasis in terms of housing, energy, clothing (shoes) and sanitary, health (treatment) programs.

Similarly, basic collective programs provide, above all, the implementation of basic individual programs, providing individuals with food, energy, housing, clothing, as well as facilitating the implementation of sanitary, medical and reproductive programs.

All the rest, besides the basic ones, both personal and collective programs that are part of ASVP will be referred to as auxiliary or additional, although their importance for the functioning and development of ASVP is fundamental.

Such programs primarily include information and organizational ones.

These groups of programs provide the structure and "technology" of working with VPs – their compilation, improvement and modification, and, finally, their implementation.

These include programs for regulating the interaction of various VPs with each other, primarily personal and collective.

It is also an information programs for the structured presentation and memorization of VPs and their elements for future use.

In addition, these are programs for analysis, correction and modification of VPs.

Specific human programs include leisure activities, which include art, sports, the structure of recreation and entertainment.

Types of Collective Programs

According to the purpose of collective programs (CP) among them are as follows:

- Consumer-oriented, ensuring the implementation of personal internal programs

- Supporting programs – providing of other programs by energy, raw materials, performers, objects, sites
- Organizational ones, ensuring interaction between programs
- Informational, containing messages and information about the conditions for the execution of programs, as well as necessary data for the correction, modification and preparation of new programs
- Defense, security

A large group of consumer collective programs are food programs. These, in particular, include the following:

- Agricultural programs producing raw materials for the food industry.
- Food industry programs that transform raw materials to the products directly used in individual food programs.
- Programs producing individual food program makers: ovens, mixers, juicers, pots, pans, dishes, etc.
- Programs that equip sites for the implementation of individual food programs by kitchen furniture and various attributes (tablecloths, oilcloths, towels, etc.).

The next group consists of programs making clothes, shoes for use in individual programs for maintaining homeostasis. It includes programs for the clothing and footwear industries, as well as agricultural and industrial programs that produce raw materials and blanks from them in the form of fabrics, leathers, etc.

Further, these are housing programs supplying individual programs with sites for their implementation. They include the manufacture of building materials and the construction of residential facilities and the surrounding infrastructure.

An important section of consumer programs is medical programs that provide and support individual programs for the treatment and recovery of the body.

Reproduction programs include obstetric aid programs, special medical programs that support the health of the mother and child, special production programs that provide solutions to household issues.

Sanitary programs include the production of hygiene products, as well as programs for vector control of diseases and support at the proper level of the residences.

Leisure programs provide recovery of a person after work, i.e., participation in a collective or individual main personal work program.

Leisure programs include entertainment (cinemas, theaters, circuses, concerts) and sports programs. In addition to the implementation of collective leisure programs that serve mass visitors, there are a large number of programs that produce objects of art, entertainment and sports to carry out the corresponding individual leisure programs.

The next section of the collective programs is supply programs that provide all the necessary supply to the other programs.

These include energy programs that produce energy, primarily electricity, and supply it to specific consumer programs.

Further, these are raw material programs that produce source material for all producing and energy programs. It uses mineral, natural and agricultural sources.

To the supplying programs, we also include those that are made for all the production programs of makers and co-makers, which in traditional terminology are called the means of production.

The creation and equipment of the sites of collective programs are engaged in the company of industrial and civil (not housing) construction.

The programs to produce components and blocks for "end product" assembly programs should also be mentioned.

Since the overwhelming share of modern collective consumer and procurement programs is complex and is carried out in the context of extensive cooperation, organizational and informational programs are crucially important.

Organizational programs that provide interaction between programs include transport and communication programs. Transport programs carry out the necessary exchange of material products and people, while the communication ones – the exchange of information. Both types of programs are supported by production programs aimed at the development and release of new modifications of both the means of transport and communications and the necessary infrastructure.

Road, air and rail transport along with water transport form the basis of transport programs.

To the traditional telephone, telegraph and radio communications in recent decades have been added mobile and Internet.

Information programs, which include radio and television broadcasts, various media on paper: newspapers, magazines, books and of course, Internet provide information about a variety of programs: the course and conditions for their implementation, about modifications and new developments of the programs.

In addition, information is provided on the current natural and social situation, the situation that should be considered in various program activities.

Finally, an important section of information programs is those that contribute to the training of individuals and their teams.

Information flows are divided into two large groups: mass, designed for a wide consumer without taking into account its program orientation, and special, which is precisely program oriented.

A very important subsystem of collective programs is creative programs of science and art, which create information that guides the community to expand the program space, develop new programs and improve the management and interaction of all the programs included in the ASVPs and the whole ASVP in general.

ALLOCATION OF COLLECTIVE PROGRAM RESULTS: ASVP MANAGEMENT AND SYNCHRONIZATION

Goods and services resulting from the implementation of the collective programs reviewed (except for the results of part of the creative programs) are consumed and used by other personal or collective programs.

To realize the allocation of goods and services produced, two mechanisms are used – the market and distribution mechanisms, implemented by the corresponding programs of the same name, or rather, by the program systems.

The market is a natural or virtual warehouse of finished products (goods and services) offered by the manufacturer to the consumer (buyer), under the conditions prevailing in the market at the time of sale and determined by the ratio of demand from the buyer and supply from the manufacturer. Those when using the market mechanism, the interaction of production and consumption programs is regulated exclusively by feedbacks between these programs in the absence of external management.

An unambiguous and universal exponent, the embodiment of the terms of sale, i.e., the value of the goods is money. You can assess the value of any product and service by this universal equivalent.

Each program for its execution requires resources in the form of manufacturers, facilities, execution sites, energy and labor. In a market system, all resources are goods and services supplied to the market by other programs. In the case of labor, this is the market offer of personal individual programs. The equivalent of all these resources is money.

Thus, money, or rather the system of financial programs that organize and regulate cash flows, is the basis for the functioning and development of the entire ASVPs when the market use for the realization of produced goods and services.

In the case of the implementation of the distribution mechanism, the consumption of goods and services is controlled by special volitional, command and administrative programs, bypassing the market mechanism.

Accordingly, in this case, the functioning and development of the ASVP as a whole is also implemented by administrative means.

A state with market governance of ASVP is denoted by the term "market", and its antipode with a strong-willed, command control – "administrative".

The ability to manage a program, a program system, and, in general, an ASVP is usually called power, which can be personal, authoritarian or collective.

Under market management, power is in the hands of a person or group with significant financial resources and managing a system of financial programs.

And with administrative management, the determining factor for those in power is the management of the system of administrative programs.

It should be noted that at present there are no countries in the world with a definitely market or administrative control system for ASVP. Mixed control systems actually work, usually with a predominance of one of them.

The structure of the ASVP control system includes control programs of various levels for both collective and personal programs.

Here we consider only individual, in our opinion, characteristic elements of this structure for ASVP with mixed control.

The control objectives of ASVP, like any other stand-alone program system, are as follows.

A tactical goal, which is aimed at maintaining the stability of the system in response to the current deflecting and unbalancing effects from both external and internal factors.

A strategic one is focused on increasing its resistance to these impacts through evolutionary development.

External factors include environmental conditions and phenomena, programs of nature representatives, hostile to the ASVP, neighboring ASVP.

Internal factors include inconsistencies between programs and their systems of different groups and segments of the community that make up this ASVP, including market fluctuations.

The following methods are used to manage programs and their systems.

First, the intensity of their implementation varies, increases or decreases, by changing the number of performers and co-performers. For example, the number of workers in production and the number of machines involved change. At the same time, working areas, volumes of purchased raw materials and components may change accordingly.

The execution of irrelevant programs is stopped, and the actual new or modified programs are launched. Accordingly, there is a constant demand for developing scientific programs that prepare modified and innovative programs, as well as creating blanks and prerequisites for them.

The observance of the regulations for the execution of programs, the so-called production discipline, as well as the interaction of the constituent programs, included in the subsystem is maintained and monitored.

In order to implement ASVP management, which is carried out by special control programs, resources are needed not only and not so much on the control programs themselves, as on the restructuring and updating of the controlled programs. In most cases, such resources are both money and the material component: executors for managed programs, their objects, new or modified sites for their implementation. For the most part, monetary resources are used in ASVP with a predominantly market management method, while direct use of material resources is characteristic of the administrative one.

The administrative management is as follows.

Plans for the achievement of program results in certain terms, instructions and regulations for the implementation of program scenarios and conditions for their implementation are formed and monitored.

The structure of the ASVP administration system correlates with the geographical structure of the given state, which has various levels: federal, regional (provincial), district and municipal. The higher the level of structure, the larger the tasks and strategic long-term goals it sets and controls.

In addition, there is a sectoral management structure for different types of goods and services produced, which is also built on a scalable hierarchical basis.

At the same time, the principle of scalable division of the industry into ministries and companies, plants and enterprises, divisions and departments is determined primarily by the convenience and efficiency of management of the corresponding program subsystems.

An important role in managing ASVP and its structural components is the coordination between programs and their systems.

Such coordination is necessary between various collective, collective and personal and separate personal programs.

Most often, the need to coordinate the implementation of individual, especially collective programs, is caused by their mutual claims for the use of resources, both financial and material.

Controversy is widespread about claims for various resources between personal programs of relatives and neighbors.

Finally, one of the most important points for the optimal existence and development of an ASVP is to minimize the contradictions between collective and personal programs. As already noted, the individual's program system consists of personal internal and external programs. The internal program primarily ensures the functioning of the organism of the individual and his family members. An external program is a component of the collective program in which an individual works and serves as a source for providing the internal program with resources.

The main contradictions between the internal and external programs that require resolution are related to the need to allocate time and energy costs between them, and thus between personal and collective programs.

Usually for coordination, special coordinating programs are used, which in the most acute and difficult to solve situations are the courts of different specializations and instances.

The competence of the territorial (geographical) authorities includes a wide range of issues.

These are issues of the functioning and development of collective programs: consumer, procurement, organizational and informational, with an emphasis on consumer programs. The relevant departments of local and federal governments deal with these issues.

Further, this is the control and regulation of the supply of personal programs with everything necessary from food to energy resources and from medical to leisure-entertainment services, carried out by the managing units.

They also include social programs: payment of pensions to the elderly and disabled, various benefits to the deprived members of the population, implemented by the departments of the pension fund and social protection of the population.

Important functions are the regulation of disputes between different programs and the maintenance of public order. This is done by an extensive system of programs of the Ministry of Internal Affairs, the Prosecutor's Office, the courts and the penal system.

Finally, it is a tax system that collects money in the form of taxes from individuals and legal entities to finance the authorities in question, i.e., control programs they run.

All the above-mentioned program directions of the ASVP of the state represent the internal subsystem. However, there are external ASVP programs that are mainly managed by federal agencies. These include primarily the relationship program with external ASVP.

The Department of Foreign Trade develops and monitors trade in goods programs with other ASVPs.

The military department develops and implements defense doctrine programs. In accordance with it, the armed forces are being created and developed, as well as programs for their use against the alleged enemy.

In addition, federal agencies are involved in managing global joint programs of this ASVP with its neighbors, primarily environmental, space and large scientific programs, for example, on thermonuclear fusion.

All of these centralized federal programs are funded by tax revenues.

In general, the control system of any modern state has the following structure.

The central part of this structure is the system for maintaining the stability of the ASVP (often called the legal system) through the formation and implementation of special programs to maintain proper and adequate relationships and interactions between all types of ASVP programs – personal and collective.

The basis of the legal system is the constitution – a set of basic principles of the relationship between collective and personal programs, between the state, administrative management system and the individuals that make up this state. Moreover, the main content of the constitution is the enumeration of the rights and obligations of the parties in the performance of their personal and collective programs.

Further, it is a set of laws, the main part of which is integrated into criminal and civil codes, that detail the rules and conditions for the interaction of various programs with each other – personal with collective and also personal and collective among themselves. Essential in the body of laws is the presentation of the principles for resolving contradictions in such interactions and the determination of the measure of responsibility and punishment for violation of the above-mentioned principles.

The Ministry of the Interior identifies law violations at implementation of both personal and collective programs, identifies the violating party called the suspected party at this stage and facilitates the appeal to the court about the law violation.

The courts carry out programs for the implementation of laws and, in accordance with them, resolve contradictions between programs that determine the measure of responsibility of the violating party and punish it.

The penal system enforces the court appointment.

The Prosecutor's Office oversees the work of the Ministry of the Interior, the courts and the penal system.

In addition to the above legal system, there is a state system governing (in varying degrees depending on the type of state) the functioning and development of sectoral collective programs – sectoral ministries.

Among them are the key ministries of finance and economics. The first provides a stable and promising state of the financial system of the state, based on the national currency. In simple terms, the Ministry of Finance primarily monitors the filling of the state treasury and thus the financial resources of the budget.

The budget is the main financial annual plan-law of the state that consists of two parts – income and expenditure. The revenue part is replenished mainly from the already mentioned taxes, as well as from the state-owned property sold in hands. The expenditure part, approximately equal to the income financing provides financial resources for such state programs as law enforcement, social, all foreign policy, including defense, and partly, in accordance with the internal political structure and the adopted national development strategy, sectoral programs.

The Ministry of Economy is engaged in developing and monitoring the strategy of economic national development, largely forming the expenditure side of the budget consumed by the selected priority sectoral areas.

The remaining thematic ministries – production ones by energy engineering, agriculture, light industry, and social ones: the pension fund, the Ministry of Health and Social Welfare ensures the functioning and development of programs in the accountable industries within the allocated part of the budget.

All the ministries listed above are united in the government, which, headed by the Prime Minister, implements the internal and external policies of the state.

Depending on the political structure of the state, domestic and foreign policy is formed by a collegial body – a parliament elected by the population or individually the head of state (most often the president), or in most cases, both together.

The participation of the population in the management of ASVP as a whole and of specific programs included in it all increases and this is due to two main circumstances.

First, the number of interrelated programs in ASVP is rapidly growing, which makes it difficult to adequately manage administrative ones, i.e., in a largely centralized way. Many issues need to be identified and resolved, as they say, on the spot.

In addition, the level of political literacy of the population is growing together with the degree of its self-organization in the cells of civil society.

Both of these factors contribute to increasing the importance of public participation in the management of both the programs themselves, and especially in identifying the contradictions between the various programs and their subsystems.

This participation is realized through a program of general elections of supreme (parliament and the president) and local governments. Parliament and the president appoint the government and the leadership of the law enforcement units and further monitor their activities. In addition, this supreme authority approves the main financial and economic regulatory document of the state – the budget and monitors its implementation. Finally, the most important function of parliament is to propose and approve laws.

The population, through political parties or other structures of civil society, monitors the work of the governing bodies chosen by them and, in the event of dissatisfaction with their activities at the next election, changes them, as a rule, to opposition representatives.

Usually such an organization of public participation in the management of the state program system is called democracy.

Nowadays, a large role in monitoring the development and interaction of ASVP programs has been performed by the mass media, both paper and electronic. Journalists who carry out the main information-gathering programs and conduct journalistic investigations of the most significant conflict situations that arise between different programs, so significantly affect the political climate in the country and abroad that the media call by the fourth power branch.

The remaining three branches include the legislative power – the parliament, the executive headed by the government and the judiciary.

It is the balance between all four branches of government in democratic countries with market economies, which is considered the key to the successful functioning and development of states, and in our consideration ASVP of this state.

As we know, the administrative management structure in the ASVP is not unique, especially in sectoral collective programs. The market system, with its mass coverage, is extremely effective and, in competition with the administrative system, has a much greater potential for the development of the programs it manages.

There is another program system that promotes adequate interaction and coordination of the programs that make up the ASVP. It is a moral system that includes ethical philosophical theories and religion.

A significant part of this system is devoted to the relationship of personal programs of different individuals and personal programs with collective programs. It is about the rights and obligations of the parties in these relationships, but not in the rigid paradigm of the established rules of the game of administrative law enforcement or mercantile market systems but using the conscious application of the principles proposed by this system.

Briefly, these principles boil down to two: prohibitions to take actions that violate or terminate the action of interacting or neighboring programs, and incentives to commit actions that contribute to the favorable functioning and development of these programs.

These systems are philosophical and religious, each, using their own set of arguments and evidence, though close to each other, explain the accuracy and adequacy of such behavior in their own way.

At the same time, they turn to both the practical experience of a person and his conscience.

As in other program systems in ASVP, the most important part of control programs is the analysis of the signals by which the control decision is made about its necessity, type of implementation (on/off, over/under), time (intensity) of action. The main analysis criteria are harmful/useful for a particular program and, in general, for the program system within which the analysis is performed.

In ASVP, such criteria are used in the analysis of mercantile, market and partially administrative management. Many administrative signals have the meaning of a direct unambiguous ban or encouragement and do not require analysis.

In the case of a moral management system, less defined criteria are used: good/bad, moral/immoral, noble/ignoble.

In essence, morality is the rules that limit damage, losses for interacting and just neighboring programs and contribute to the success, profit of these programs, despite own possible losses.

Unlike mercantile mutually beneficial, concrete, quickly realized relationships, moral relations imply a return of positive in a longer and indefinite period and possibly from another or future partner, a program neighbor.

In the moral relations, there is no direct experimental incentive like in the mercantile ones. The moral incentive is an innate and educated conscience.

We can say that the ASVPs consist of subsystems of personal and collective programs, managed and regulated by three types of regulatory programs: administrative, mercantile and moral.

INFORMATION IN VOLITIONAL PROGRAMS

The information component of VPs largely inherits the principles laid down and developed during the previous evolutionary stages of unicellular and MOs.

This is the network organization of the information structure, and its division into two parts – instructive, reproducing and signaling, controlling, adaptive.

All the same informational signal marks, which are the result of special informational programs, provide network interconnections between programs.

On the basis of common basic information principles, evolution has created in humans a superstructure of VPs. Moreover, the entire IS of higher multicellular, for example, primates, exists and works with minimal changes in humans in the subsystem of internal programs and in the "intuitive" part of external programs.

In unicellular and MO information support of the regulated reproducibility and adequate adaptability of the program system is provided in an automatically, innate "hardware" way. Unlike human, who while working with VPs intelligently creates new program elements, runs a program, evaluates its result and, if necessary, modifies it.

Naturally, for this to happen, a person had to master and develop all the informational principles inherent to the program cellular organizing of all living things.

Already in the very early period of the existence of Homo Programming, the formation of collective information programs was to begin, helping to capture positive experiences, as well as modify and create new programs. This may be the development of criteria for assessing the quality of makers, for example, stone cutting tools, or methods of compiling programs for maker manufacturing.

Such informational finds as well as all the effector programs of hunting, cooking, making clothes and building housing were collectively tested and memorized in the community memory.

The basis of a first collective IS was oral language. The emergence of a written language served as a huge leap in this area, as a reliable base has emerged both for creating a collective program library (PL) and for editing and creating new VPs.

Nowadays, the role of collective information programs has grown immeasurably, and the average social person, as they say, cannot take a step without collective information tools or individual information devices created by numerous corporate teams.

The main development directions of information technologies are storage devices, means of converting information patterns and arrays, means of exchanging information data between information subjects.

Next, we briefly consider some informational aspects of VPs.

Memory

An essential basic element of volitional programming is memory. Unlike the memory of other multicellular, including higher ones, human memory (HM) has fundamental features.

So, in addition to sensory input information of control neurons (n-deciders) and effector output information of the n-deciders, HM also contains a linguistic symbolic component that serves to indicate both sensory and effector information in image pattern representation.

As already noted, the language system is built in accordance with the system of programs and includes all the information associated with them. These are the elements: makers, objects, actions, results, sites, scripts; characteristics: time and speed of execution, efficiency; fulfillment conditions: the presence of prohibitions on the implementation or its suspension, a favorable situation for implementation, etc.

We assign all this program information, both image and symbolic ones, to the PL.

However, there is still a significant section of memory that contains preliminary, overview, general information about the environment and the world around a person. This information is necessary for a person to correct and modify existing programs, create new ones and, which is very important, for the coordination and proper interaction of their personal programs with their neighbors. Let's call this section the HM auxiliary library (AL). Of course, it also consists of closely related information components – figurative and symbolic, linguistic.

So, the basis for working with programs of all kinds: the regulated execution of the memorized program, its variable execution and modification of the program, also including the preparation of a new program is the PL.

It includes the following:

- Catalog of programs by their goals, objectives and results
- Program scripts with a sequence of actions and steps
- Structure of composite programs with sequentially and/or concurrently executed component programs; control schemes for component programs and their synchronization
- Performers of programs with signs, causal connections and an indication of their objects, actions performed with these objects and the results obtained
- Objects with signs and causal connections
- Types and characteristics of actions
- Descriptions of program execution sites and their features
- Conditions for the implementation of programs, including bans and permits for their implementation
- And so on

Overview, orienting information of the AL is formed by topics:

- Surrounding "geographical" environment with the possibility of expanding the horizon – apartment, house, district, city, country, land and even space
- Groups of programs being executed or intended to be executed
  - Consumer programs
  - Work programs – participation in collective programs
- Regulatory information
  - Law information
  - Financial information
  - Religious information
- Interests
  - Sport
  - Travel
  - Art
- And so on

These topics are constantly updated and replenished in accordance with regularly received news:

- In the setting – domestic, political and scientific news
- In consumer programs – updating the range and price list of goods, new consumer technologies and recipes
- In work programs – new jobs, new technologies of collective programs
- Actualization of the administrative, market – financial and religious spheres of the life regulation of the individual
- New information in the sections on "Interests"
- And so on

Thematic information of the AL is formed in the form of descriptions that consist of scenes and situations, as well as the objects that constitute them and the phenomena and events occurring in them and with them, leading to certain results. At the same time, it can be seen that even the most general and entertainment "interesting" information is structured according to schemes close to program scenarios.

More specific regulatory and coordinating information is presented in accordance with the specifics of these sections – rules, instructions, regulations and recommendations.

So far we have considered the organization of individual brain memory, which was the only basis of volitional programming until the invention of writing. With the advent of written fixation of information, artificial memory structures arise, first in the form of books and other written materials, and then with the use of electronic media having a similar organization, since it is oriented toward interaction with natural memory.

The proportion and importance of artificial memory in volitional programming and especially in collective programs is growing very rapidly. Currently, natural memory is mainly used for orientation in the structures of artificial memory, and not for actually working with primary information.

All types of work with programs are performed by accessing the memory, reading relevant information from it, reading this information, selecting the actual elements from the read section and further operations with these elements.

The materials of the AL are primarily designed to facilitate effective work with the programs.

In this sense, the AL is additional to the program.

In particular, the "geographic" content of the AL is used to clarify the conditions for the execution of a particular program, which often leads to the replacement of its performer and/or object, and possibly to a noticeable variation of the scenario. For example, when traveling to another city or, moreover, another country, it is desirable, and often necessary, to collect information about the climate, housing, transportation and food, adding to and arranging the relevant thematic sections of the AL in order to use their, and possibly modify personal home programs.

An important place in the AL is occupied by materials about neighboring, joint, parallel, partner programs that make up the program background, the landscape on which this individual acts. Information about these programs is partial and contains basic information about their goals, tasks, stages, performers, objects, websites, resources and their sources without a detailed scenario and execution calendar.

Modification of individual programs taking into account the background program situation and, if necessary, the development and implementation of special programs of coordination and counteraction are implemented taking into account the laws, regulations, rules of all three types of control of the ASVP: administrative, mercantile (market) and moral.

To prepare materials of the AL for use in real situations, it is necessary, as already noted, in addition to their constant updating and replenishment, to work on their design and structuring according to signs, causal relationships, temporary realizations (diagrams, graphs), importance and priorities, etc.

Collection, Formation and Processing of Information

Of great importance for working with personal and collective programs is the collection of information about all program elements and the programs themselves, as well as methods of their evaluation and decision-making, etc., which are necessary primarily for updating libraries as program and auxiliary. The rate of collection of such information largely determines the success of individuals and communities.

For data collection, specialized observational programs are used, the makers of which are sensors of both the person and the artificial ones, designed and manufactured to solve specific local problems.

Remember that the observation programs for the situation, both internal and external, are the main stage in the regulation of the program system in all organisms.

In addition, information is constantly being collected about the participants, members of the ASVP, to monitor their activities and the status of ASVP, in particular, the situation with the relationship, the interaction of various personal and collective programs.

The effectiveness of using the collected information is largely determined by its adequate formation and processing, as well as availability.

No wonder that computer tools for creating large amounts of data, as well as Internet search engines, have been developing intensively lately.

Information Transfer – Communication

The role of programs that inform the public about the situation in the ASVP as a whole and in local social communities using various means of communication was already mentioned above when considering the system of collective programs.

It should also be noted that the intensive development of means of communication and the exchange of primarily current information between individual and collective members of the ASVP system, continuing throughout the history of mankind. This is completely understandable, since communication is the main means of organizing an ASVP programs and its information network. The range of modern means of communication is huge – from communication between members of the group of workers, performing a common production program to the exchange of data from the participants of the space station with the ground.

The presence of individual means of communication with the ability to connect to a single Internet information network for almost the entire population of the world means the unification of mankind into a single program-information structure – a global ASVP, developing with an extremely high historical pace.

STRUCTURE OF VOLITIONAL PROGRAM MAKERS

The structure of the makers of modern ASVP is very extensive in accordance with the development of personal and collective VPs and the network of their relationships.

Despite the constant intensive growth in the number of artificial, man-made helpers of natural human makers – the hands and the brain, there is still a noticeable range of programs executed with minimal "improvised" means. This circle includes both the simplest everyday programs: washing, dressing, food and those that can be attributed to the highest achievements of human activity – art and theoretical science.

But, as for production programs in high-tech industries, here the nomenclature of the makers and the structure of their interaction are very saturated.

For example, a modern processing machine is a complex task maker, which includes sensor, analytical (computer) and mechanical (effector) block makers, which, in turn, consist of power supplies of local devices, the sensors themselves, processors, electric motors, etc. Further, the hierarchical maker structure continues at ever lower levels until it reaches the elementary details.

HISTORY OF MANKIND AS EVOLUTION OF VOLITIONAL PROGRAMS

One of the evolution goals of all autonomous program systems (ASPs), including ASVP, is to increase the resistance to external and internal influences that deviate the system from its normal state.

The potential of a program's response to deflecting unbalancing effects is determined by the level of material and energy resources and the degree of perfection of program management tools and their interaction, and consistency, and evolution consists in increasing this level and degree of management perfection.

The increase in resource saturation of programs occurs both from the side of growth in the supply of resources, and as a result of their more economical use by the program.

The evolution of personal and collective programs is carried out in concert, although with its own characteristics.

The tasks of the evolution of personal internal individual programs are to increase the sustainability of systems for maintaining homeostasis by improving clothing and footwear, housing and household appliances and objects. In order to increase the body's resistance to diseases and higher physical and mental stress, food products, means and methods of their preparation and storage become more diverse and effective. For the same purpose, medical and sanitary programs are being improved.

The development of consumer personal programs inevitably leads to the evolution of collective programs providing these programs. Accordingly, the consumption of the entire program system with energy in all its forms is increasing, which requires intensification and improvement of energy programs.

With the evolutionary development, expansion and complication of personal and collective programs working for them, organizational and informational programs are becoming increasingly important. The means of transport and communications are becoming more and more intensively improved. The role of information programs necessary for the identification and resolution of contradictions arising between increasingly numerous and diverse programs is rapidly increasing.

Thus, the evolution driven by personal programs embraces the entire autonomous system of VPs.

The essence of the evolution of any program comes down to the evolution of its components: the performer, the object, the site and the script, which in the spirit of time can be called technology.

The most important examples of the early stages of the evolution of VPs are the emergence of such new performers as a spear, a bow and traps in hunting programs; technologies for using fire in cooking programs and maintaining temperature homeostasis; knives for cutting animals and skinning, as well as needles and the technology itself sewing clothes and shoes from animal skins in programs for maintaining homeostasis, as well as moving; the technology of cultivation of nutrient plants near the dwelling for use in the food program; technologies for taming animals for use in food and transport programs; technology for building houses to use them in programs for maintaining temperature homeostasis, in protective programs and in equipping sites for many household programs: cooking and eating food, making clothes and shoes, making hunting and household tools.

All of the above programs could not arise and improve without parallel development of the level of interaction and coherence of personal programs when they are combined into increasingly complex collective programs, which required improving organizational and information programs within the capabilities of a developing oral language.

The first social structure emerged – the communal-clan system – a prototype of an autonomous system of VPs in which its functional elements of management and regulation already existed: imperiously administrative through the family and clan hierarchy, mercantile through the system of mutual services, moral through the system of totem bans.

Further key evolutionary historical stages are also associated with noticeable and rapid improvements in program subsystems of ASVP in all directions, some of which are discussed below.

Basic Consumer Programs

- Cultivation of grain plants and mastering the culture of production and consumption of bread products

- Mastering animal husbandry and food technology for food programs and making clothes and shoes
- Mastering textile and sewing technology

Supply, Procurement Programs

- Development of natural materials to produce program performers from them (mostly): lead, copper, iron
- Mastering synthetic materials: bronze, glass, cast iron, steel, including alloyed with various chemical elements, plastics, composite materials
- Development of combustible materials, types of fuel: coal, oil, gas

Energy Programs

- Development of new energy sources: water, wind, steam, electric, nuclear

Transport Programs

- The invention of wheeled carts
- Use of riding, pack and cart animals
- The invention of sailing water transport
- The invention of steam, internal combustion, jet, rocket engines
- The invention of rail transport
- The invention and development of road transport
- Invention and development of aviation transport
- Inventing and mastering means for space flight

Information Programs

- Development of means of exchange, transfer of information
  - The appearance of messengers, couriers, heralds
  - The appearance of the bells
  - The appearance of sea and land signals
  - The appearance of mail
  - The invention and development of the telegraph
  - The invention and mastering a telephone
  - The invention and development of radio communications
  - The emergence of satellite communications
  - The appearance of the Internet and e-mail
  - The emergence of mobile communications
- Development of methods for recording and storing information
  - Written language
  - Mathematical language
  - Graphic language
  - Typographic method of recording and reproducing information

- Inventions and improvement of storage media
  - Clay tablets
  - Papyrus
  - Paper
  - Electronic storage:
    - Magnetic disks
    - Microchips
    - Optical discs
- Development and improvement of programs for the presentation, processing, analysis and processing of information
  - The emergence of programs compiling recipes, instructions and rules
  - The emergence of programs compiling reports for rulers and leaders
  - The emergence of programs compiling biographies
  - The emergence of programs of compiling codes of laws and explanations for the proceedings of program contradictions
  - The emergence of programs for the preparation of textbooks and materials
  - The emergence of speech logic, stimulated by judicial and pedagogical practices
  - The emergence of geometry based on a graphic, pictorial language and at the same time a mathematical system of evidence
  - The emergence of digital mathematics
- Development of means and methods of disseminating information
  - The emergence of centers for information accumulation (elders) and students in them to whom this information is transmitted
  - The emergence of pedagogical centers
  - Typography
  - The emergence of schools and universities
  - The emergence of newspapers and magazines

Stages of Administrative Management and Regulatory Programs

- Structure of primitive communal hierarchical management
- The first state geographic structures with sole management
- State structures with the division of the population into slaves and free citizens
- Feudal states with vassal substructures
- Capitalist states
- Socialist states
- Mixed-type states – capitalist with a large share of socialist institutions

Development of Market Regulatory Programs

- The formation of the exchange barter market of goods and services
- The emergence of a single equivalent of goods and services in the form of portions of precious metals

- The emergence of retail and wholesale markets
- The emergence of the money market
- The emergence of commodity and money exchanges
- The emergence of a centralized state financial system with treasury and taxes

Moral Regulatory Programs

- The emergence of the first religious pagan systems to appeal to the highest, omnipotent power, unknown, but supposed and able to control mainly the external stochastic processes necessary in a predictable and dosed way, for example, to stop the storm, to cause a good (not very strong) rain. In other words, convert the stochastic process into a predictable forced program.
- The emergence of monotheistic religions with a clear system of basic conditions for the implementation of personal programs and coordinated participation in collective programs and prescribed elements of universal education, mastering these conditions by methods of punishment and encouragement.

Military Programs

- Invention of the striking weapon – executors and co-contractors of military programs: combat bow, sling; cold steel arms: daggers, swords, spears, halberds; gunpowder and firearms; automatic small arms; armored vehicles; combat aircraft; combat missiles; atomic and thermonuclear weapons; space weapons; laser weapons
- Creation of appropriate programs of warfare against the enemy for each new type of weapon

It is possible with a certain degree of conditionality to distinguish two directions of evolution of ASVP: the main direction is the improvement and generation of new VPs and auxiliary is the accumulation of all types of resources for the implementation of the main direction.

The methods of implementing the auxiliary direction can be both internal and external.

The internal types include various types of savings, for example, centralized in the state treasury, local in state subjects, industry, in associations, companies running collective programs and, finally, personal ones.

Among the external methods can be noted foreign trade and geographical expansion by the conquest war.

The benefits derived from evolution are disseminated and consolidated by exchanging information using both administrative and market mechanisms, both within and outside the ASVP.

Thus, with the development of information, coherent and transport programs and the globalization of market and administrative control programs, the pace of evolution of ASVP increases dramatically.

On the whole, the pace of evolution of the ASVP already since its inception several tens of thousands of years ago exceeded the speed of classical gene evolution by several orders of magnitude.

Practically unchanged from the time of the Cro-Magnon era on the genome, modern man every few decades finds himself in a fundamentally new structure of ASVP.

One of the basic conditions for a long-term effective evolution is the connection of this ASVP with its neighbors, preferably more developed ASVP, in the field of trade, exchange of labor resources, and most importantly in obtaining information on the next evolutionary steps in various program subsystems. The conglomerates of ASVP arising in the case of such connections are usually called civilizations.

There are cases of quite a long successful evolution of one closed, but large state, for example, China.

But small closed ASVPs, to which all tribal formations of Africa, Asia, America and Australia can be attributed, drastically lag behind the other "civilized", evolving states.

Nowadays, due to the rapid development of coherent, transport and information programs, the isolation of individual ASVPs becomes impossible and the nature of evolution assumes a general, global, planetary character.

The involvement of backward ASVPs in the modern evolutionary process, as a rule, is quite successful, which once again testifies to the proximity of the genomic capabilities of all races, peoples, nationalities and even tribal communities without exception.

Similarly, as the genetic evolution, ASVP evolution at any particular stage affects only certain, relevant program subsystems and not all of them at the same time, since this would require too many resources.

Any program system spends the bulk of resources on reproducing a regular state condition, and not on evolution. However, in the ASVP, as the evolution accelerates, the share of resources required for it is constantly increasing and in fact, it limits its even higher rates.

SCIENTIFIC STAGE IN EVOLUTION OF VOLITIONAL PROGRAMS

As already noted, the history of mankind is the evolution of VPs, or rather, autonomous systems of these programs of ACVP, and the main direction of this evolution is the improvement and generation of new VPs. Up until the end of the 16th century, significant evolutionary events that sharply accelerate it occurred more or less randomly as resources accumulated in the most "advanced" ASVPs.

And it was only from the 17th century that evolution began to actively develop a special mechanism for its acceleration – science.

At the same time, the ASVP control programs constantly increase the share of resources allocated to this subsystem, which corresponds to the general tendencies of accelerating the evolution of individual ASVP in the situation of their intensifying competition.

Science is a specialized information program subsystem of ASVP, which has the goal to obtain, create, organize and systematize new information for the development of all ASVP programs, including scientific programs themselves.

Applied and Fundamental Science

To date, the scientific program subsystem, further traditional science, has a widely developed structure, largely corresponding to the structure of ASVP and determining the modernization of all consumer programs, including food, household, construction, medical and entertainment programs, as well as providing them by the programs: production, supply, energy, etc., and also, which is especially important at the present stage, information.

But this is only a part of science, although it is the most voluminous and costly, which, according to our terminology, supplements the PL of memory, and it has the generally accepted name of applied science.

Its other fundamental part prepares the scope of work for applied science, expanding and refining the AL.

Scientific Methodology

Some elements of the methodology of fundamental science on the example of physics will be considered further.

We chose physics, in particular, because it studies the basic laws of the non-living world and has developed the most distinct methodology inherent in the exact natural sciences.

The basis of physics is experiment. Let us try to formulate and justify its definition from the position of the program approach.

An experiment is a recording of the reproducible reaction of an inanimate object to a regulated program of experimental exposure.

Indeed, a competent classical physical experiment requires the following conditions to be fulfilled:

- Selecting an object that allows you to unambiguously observe its investigated parameters.
- The choice of a certain modality and regulated in space and time impact on the object.
- Exclusion or, at least, suppression of all influences and the phenomena caused by them that are not related to the studied ones.

The experimenter unambiguously influences the object in a certain specified place of space and at a certain time with the help of his own means (makers) predictably and regulated and observes the result of this influence on the object. In this case, all extraneous parasitic processes and phenomena, including the stochastic behavior of the object itself, are excluded as much as possible.

In other words, the experimenter examines how the given maker (means of influence) will execute the program in relation to the object and to what results this will lead. The result of this research is recorded in the form of a regularity (law) of the object's behavior under a given "clean" effect on it of a certain maker.

Clearing the experimental program allows you to isolate the predictable, essentially programmatic manifestation of an inanimate object. The experimental program necessarily contains the verification stage of such a cleaning, by fixing the reproducibility of the experiment results at its repetition. Only a pure experiment can be reproducible. A typical example of the purification of the experiment is Galileo's experiments on the study of the fall of objects with different ratios of weight and air resistance in an evacuated tube. By eliminating the impact on the studied objects of air resistance, he showed that all objects, for example, a feather and a stone, fall with the same acceleration.

A new physical result becomes generally accepted when it is reproduced in the same regulated experimental conditions by several independent researchers.

The physical result is usually considered to be a reproducible change in the property, state of the object under study when exposed to other objects or, on the contrary, registration of changes in the objects affected by experiment.

For example, in one case, the electrical and magnetic properties of various objects were investigated by experimental programs in which an electric object, for example, a current coil acting as a maker, acted on a magnetic (e.g., magnetic needle) serving as an object. And in another, on the contrary, the permanent magnet, being a maker, acted to the same frame with current, which in this experiment became an object.

In general, in the program system, makers and objects change places in different programs – a maker in one is an object in the other and vice versa. A good example is when a protein maker becomes an object of an anti-stochastic decomposition program. Not to mention that all potential makers are objects up to the last stage, the step of creating, activating and transporting them to the site programs. For example, a ready-made protein remains the object of a transport program that delivers it to the site of its upcoming work – the execution of a program in which it is a maker.

Naturally, the same situation is observed in the system of VPs, including the scientific programs under consideration. Therefore, when we say the "object" of research and/or observation, we mean the possibility of its further assignment both as the makers and as the objects of the corresponding programs.

Thus, the studied objects of inanimate nature become makers and/or objects of the experimental program, and then they are used in real innovative programs, first in applied science and then in the practice of the main existing programs under the same conditions. Thus, the objects of the inanimate stochastic world and/or their properties are mastered and assimilated by the program system.

It should be noted that the same, in essence, but less developed approach to the use of non-living objects in programs was used before the advent of science from the earliest stages of the evolution of ASVP.

It can be said, using the initial ideas of the present work, that physics forms a programmatic perception of stochastic processes of the non-living world, using a regulated experiment.

And in general, the scientific understanding of the inanimate (non-programmed, stochastic) world is a set of reproducible reactions of its objects to experimental clearly regulated exposure programs.

At the same time, the rules of the program of physical experiment, based on the concepts: time, space, causality, etc., are attributed to the stochastic object of study, and then to the entire stochastic non-living world.

One of the main characteristics of the object observation is its movement. It is not by chance that the basic section of physics is mechanics, which largely studies the laws of motion of bodies, under the action of force. The world basic concepts of time, space and causality originated in this section of physics are successfully applied in its other sections, as well as in other sciences.

Use of Space and Time Concepts

Let's try to consider a schematic scenario of the program for monitoring the movement of an object, for example, a satellite in the night sky.

First, we choose this moving object among motionless, for example, stars. We do this of scanning by the eyes of the neighboring areas of the sky and observing and fixing each of these areas on the visual analyzer at least twice. In this case, the result of the first fixation is memorized, and the result of the second is compared with the first. As a result, in the area with a moving object, we find the difference between the two fixations in changing its position relative to neighboring objects, which did not change their position relative to each other. Moreover, the moving object in these first observations is detected with the greatest probability not in the center of the visual zone with the best spatial resolution.

Then we begin to determine the vector velocity of the satellite in both direction and magnitude.

First, we place the image of the object in the center of the visual zone and as it moves, we maintain this position due to eye movement and, if necessary, neck movement.

To determine the direction of motion, we carry out the same pair of fixations and the subsequent comparison of the images of the object together with fixed objects. Analyzing the results of the comparison, we fix the distances between the satellite and the stars that have increased the most, and those that have decreased the most. The direction of the satellite's motion, we determine – from those stars from which it has retired to those it has approached.

Next, we determine the speed of movement. To do this, we observe changes in the distance between the satellite and, for example, the star of rapprochement by double-fixing the image of these objects with the time interval between these fixations, determined by the number of the repetitive, periodic process, for example, the pendulum clock or own heart rate.

Thus, we simultaneously and synchronously make two observations. We observe the satellite and the time and we compare their results.

After the already known procedure for comparing the current second image with the first one remembered, by subtracting the second distance from the first we get the satellite position change, i.e., distance traveled for a specific time between observations. To obtain the value of the velocity, independent of the time between observations, we must divide the determined difference distance on the number of realizations of the periodic process observed during this, and then we will get the speed as the distance traveled between the realizations of the periodic process. Thus, we get the classical definition of speed, as the distance traveled per unit of time.

So, we emphasize once again that the task observation program consists of two programs executed in parallel: the first to register changes in the object's characteristics and the second to track the progress of the periodic clock program. Both are performed with the active participation of memory.

Aspects of Time Concepts in Program-Information Approaches

First, we'll try to clarify the term "program flow".

By definition, a program is a sequence of regulated steps, actions performed at a given pace of implementation, the sum of which leads to a result at a given, also regulated, predetermined time. Those introducing the definition of a program, we mean its temporary implementation, or, in other words, we present it as a regulated stream of sequential actions by the maker and, as a result of them, changes to the object.

The program system of any organism, including, of course, humans, is a combination of such program flows.

One of the main tasks of managing a program system is to synchronize these streams, in particular, to implement programs at the highest level of the system, for example, the cell cycle program. Special proteins – cyclins, play a crucial role in this synchronization in the cell. In certain phases of the cell cycle, they give signals for slowing down or completely stopping some programs and starting or speeding up others, thereby regulating the speed of streams of individual programs and their system.

The periodic appearance of cyclins in conjunction with their signals, which use cellular control programs to synchronize and control the speed of program flows, is a direct analogue of clocks used by humans for the same purpose in controlling VPs.

The basis of the watch mechanism is most often the process that occurs with inanimate objects, for example, these are periodic oscillations of the pendulum. In the watch device, this process is converted into an adjustable program with a fixed reproducible duration, at the ability to monitor the progress of this program, i.e., fixing the current time, for example, using the dial and hands.

Our life is a combination of VP flows, and we need a special program for synchronizing them, similar to the cellular cyclin program.

Such a program, or rather the program subsystem, is the watch industry of VPs, both collective and individual.

Collective watch programs mainly develop, improve, manufacture and repair watches.

Individuals in their watch programs mainly carry out observations of these watches, combining, synchronizing with this observation, the performance of other current programs, not forgetting, of course, to start the watch on time.

The watch monitoring program is carried out in several versions.

Common individual program is as follows. When drawing up daily and more detailed hourly and minute programs, a schedule is created for the execution of stages and the synchronization between different programs according to the time zone. Then, as they are executed, the position of the watch hands is recorded, followed immediately by a comparison with the currently stored position of the hands, which is supposed to follow the schedule of one or several programs being executed.

In accordance with the result of this comparison, deceleration or acceleration of stages, or programs in general, is carried out.

Another version of the watch program is used in conjunction with programs for monitoring moving objects, as is the case in the example of determining the speed of a satellite that we have analyzed.

Many observation and research programs are carried out for predicting the further behavior of the object, in order to use the information obtained to supplement and improve the program and, for the most part, AL. And in all these programs, the key is the synchronous use of a clock program, which in essence is the source, the basis of the concept of time. It is thanks to the widespread implementation of this time, time program with the same hours selected by agreement that the necessary synchronization of all programs performed by people is carried out.

For the same purpose of replenishing the AL, a number of observations and studies, for example, in geography, geology, evolutionary biology and others, are aimed at restoring the course of programs or processes that have occurred in the past.

In this case, a convincing reconstruction of the sequence of events can be made only if there is an hour process accompanying these events with guaranteed duration. An example of such a process is the decay of a series of radioactive elements, successfully used to synchronize and determine the sequence of past events in the mentioned areas of knowledge. At the same time, just as in the examples of using watch programs considered above, it is important that the synchronization of the event being studied coincides with the time mark.

The use of geological layer methods for such synchronization, in which it is important that the level of localization of the investigated event coincides with the layer from which the synchronizing radioactive material is extracted, can be called into question if the normal sequence of horizontal layers is violated.

In all the examples considered, the use of a special time synchronization program gives rise to the concept of time as a parameter that characterizes all the programs that are executed and observed, as well as the processes that are observed and used in the programs.

In other words, time is a purely programmatic concept, used, in particular, to describe inanimate nature.

All organisms exist, live at their own pace, characteristic of this program system, for most of the programs executed, at least external ones. For example, snail and hummingbird performance are very far apart. For a faster system, slow movements are easily overcome, weakly compete. For example, a person is not able to catch a bird or a fly on the fly, and fast birds – which can easily catch flying midges.

Thus, each program system has its own temporary program. In the living world as a whole, there is no single time for different species, with the exception of the general daily and annual cycles, according to which a limited number of internal programs are synchronized.

All organisms have a system of synchronizing biorhythms in both internal and external programs.

For a modern person, external VPs use a unified system of temporary programs, in which substantial attention is paid to synchronizing specific, programs with each other. As a result, the human community creates an idea of a common single world time, which is given the status of a basic concept.

Stochasticity at Inanimate Nature

When we talk about the stochasticity of the non-living world, we mean the randomness, the unpredictability of the implementation of certain events and processes in it, as opposed to regulated, defined, predictable programs. An event or process occurring with an inanimate object is the result of a random combination of effects on it of other non-living objects. By the way, the very allocation of a separate object for observation and the subsequent tracking of its behavior and the classification of this behavior are possible only with the help of the experimental programs considered above. This situation is clearly demonstrated and described in gas thermodynamics, where the average time between events is much less than the observation time.

If we take the ratio of these times as a parameter of stochasticity, we can roughly divide all the events and processes of the inanimate world into stochastic manifestations observed with a stochastic coefficient $Kc \ll 1$ and non-stochastic manifestations observed with $Kc \gg 1$. An example of the first is the events that occur with gas molecules, and comets flying in deep space belong to the second.

It is clear that non-observables as stochastic events of the second type do not cease to be stochastic in nature, i.e., unpredictable, unregulated and non-programmatic in the meaning of the term we use. The predictability of the observed piece of the comet's trajectory means that the number of events by the change of its trajectory during the observation is not more than one and it should occur under the influence of a known object, for example, the sun, i.e., in the conditions of the same program experiment. But what we could not predict was how the comet got on this trajectory of approaching the observer, since this happened as a result of many stochastic influences.

A very important kind of process of inanimate nature is periodic processes, which are also stochastic due to the randomness of their formation. For example, the planetary systems formed due to the stochastic gathering of small cosmic particles around larger already formed cosmic bodies by this time. And when this process of completion of the planetary system ended, it entered the stage of small and rare external influences in the form of a relatively stable complex of planets moving along closed trajectories. Observed by us people and all other program systems, since the beginning of their existence on earth, this stable state of the solar planetary system has essentially served as the basis for this life not only as a source of material and energy resources and acceptable homeostatic conditions, but also as a stable reproduction of annual and daily rhythms, which seems to be a very important factor for the necessary reproduction of the life cycles of organismic program systems.

So, the predictability or even the relative stability of the processes of the non-living world that is local on the observation time does not contradict their fundamental stochasticity, but is successfully used by the living program world for tactical stabilization and strategic evolution.

An important manifestation of randomness, unpredictability of events and processes of inanimate nature over time periods, including a sufficient number of events for their stochasticity, is the inevitable irreversible degradation of such quasi-stable ordered states as the considered example of oscillatory processes.

In other words, an inanimate system can accidentally go into a relatively ordered energetically favorable and relatively stable, for example, vibrational state, for a time determined by a weak but always existing stochastic effect on it and characterized by damping. However, other random processes cannot compensate for this damping decrement, let alone switch the system to a higher level of ordering, and therefore, in accordance with the second law of thermodynamics, it is doomed to degradation.

Therefore, an evolutionary process that is not only accompanied by compensation for the stochastic degradation of an ordered system, but also leads to an increase in this order and the stability of the system similar to the one we see in the living program world, is impossible.

In this regard, the term evolution of stars or galaxies is incorrect.

The approach to the processes of inanimate nature itself may be somewhat adjusted in light of the stated program concept.

Combating Stochastic Degradation in Program Systems

The opposition of stochastic degradation in program systems occurs in several directions.

One of the basic principles is the availability of the stage of updating the organismic system of programs at the maker level, which is implemented by several program subsystems.

There is an update for the makers. Repeatedly, in the course of a single cell cycle, spent makers are decomposed, followed by the removal of their fragments from a cell and the creation of new makers according to gene instructions.

In addition, each cycle after cell division, there is a general update of all the performers.

Further, stochastic defects arising in DNA instructions – DNA molecules, are eliminated, healed by special reparation programs.

In MOs, a developed complex, ramified system of programs, subject to degrading stochastic influences during active interaction with the external environment, is destroyed completely when the body dies.

The convolution of the organism program, which is located in separate male and female reproductive cells, and then combined in a fertilized egg cell, is preserved and inherited.

Reproduction programs and subsequent unfolding of convolution into the complete organismic system occur at the young age of the mother and father with minimal accumulation of organismic degradation defects that nevertheless accumulate despite the considered measures to combat them inside each organism.

The aforementioned areas of counteraction stochastic degradation can be classified as operational, occurring during the life of a single organism.

In the presence of a sufficiently large species population of organisms that really differ from each other in degree of degradation, primarily in genes, generational change leads to the removal of individuals with a high degree of degradation from populations.

This occurs in accordance with the described evolutionary mechanisms, one of which, for example, is the predominant reproduction of healthier and stronger organisms. In the case of a strong degree of degradation of a fatal nature, the death of the body occurs in the first generation.

The considered evolutionary direction of counteracting degradation can be attributed to the strategic one. It plays a significant role in the stabilization and improvement of the species program system and requires for its implementation a clear selection and isolation of a sufficiently large population of identical species organisms by the mechanism of reproduction.

Thus, the identified species communities, and not a multitude of small intermediate interspecific groups of organisms, whose absence is used as one of the arguments of criticism of Darwin's theory, are evolutionarily stable and promising.

The role of evolution in counteracting stochastic degradation is not only to cleanse the species population from defects in the program system caused by it, but also to improve the body's anti-degradation programs.

ABOUT LEARNING

In ASVP, learning mainly comes down to the replenishment, perception and assimilation of materials from both libraries – program and support ones. Mastering implies the possibility of practical realization of perceived programs and ways of working with them.

Usually, the mastering of a PL containing specific programs, both personal and collective, is called the term learning itself, and filling AL and working with it is called education.

The most important task of education is acquaintance with the structure of the ASVP, primarily with the subsystem of management and regulatory programs of all three types: administrative, market-mercantile and moral. This is necessary for the formation of a full-fledged socialized member of the ASVP.

As a second (in order, and not in importance), the key pedagogical task can be called mastering the technology of working with VPs, essentially technology of thinking. This means first: the ability to select and set a task, to compose a script of a general task program, to develop a scheme of component programs that form a task program, their results, as well as makers of different levels. Next is the study of all other program elements: objects, sites and resources. Great attention is paid to the development of methods for evaluating both the program elements and the programs themselves, from the point of view of time and resource costs and efficiency. The results of such assessments should be used to select programs and their elements from the options considered.

It seems to us that the formulation and implementation of the two listed educational tasks could contribute to the effectiveness of the functioning and development of ASVP.

Training in individual programs at the primary level in childhood usually takes place in the family and then proceeds in the manner of personal self-study using personal experience, observations and using various primarily mass information sources.

ABOUT THE WORLD PICTURE

What we observe is not nature itself, but nature exposed to our method of questioning.

Werner Heisenberg

Picture of the world, i.e., information about the environment in which an organism, its self-contained autonomous program system executes its most important external programs, in particular, nutrition and the avoidance of dangerous factors, is inherent in any organism.

For example, for bacteria, this is primarily spatial information about food sources and substances harmful to the body, which is acquired and used when performing chemotaxis movement programs.

For MOs with a much more developed and complex system of external programs, information about the external environment becomes much more significant.

Finally, for a human personally and for program communities, of which he is a member, the picture of the world is fundamentally important in all manifestations of VP activity both for executing and replenishing the library of ready-made programs, and for creating and editing an auxiliary informative library.

At the same time, along with the expansion of the areas forming this picture, their deepening and refinement take place. Sometimes it is necessary to correct its structure.

It seems to us that the proposed information and program concept of the living allow us to suggest some elements of a correction.

If we accept the proposed concept of a program structure for a living part of the world, then, creating an overall picture of the world, we have to take into account the following circumstances.

The fundamental difference between an orderly, predictable, reproducible and, at the same time, an evolutionarily developing living world from a stochastic non-living one leads to the fundamental irreducibility of the first, to a certain, higher phase of development of the second. The living world uses the non-living for its functioning and development but cannot come out of it. The genomes of organisms reproduce always from each other, although the organisms that they produce are mortal and pass into the inanimate world.

The essential difference between the living and non-living is a fundamentally different energy balance in the implementation of programs and the implementation of stochastic processes. When executing programs, most of the energy coming from outside is absorbed, spent on obtaining the result, and only a small part of it dissipates during the program or can be returned to the environment, for example, at the burning of highly ordered organic structures.

The living world constantly consumes energy.

Whereas in the processes of the non-living world, taking into account the dissipative part, the law of energy conservation is always observed.

In the inanimate world, energy passes from one form to another and dissipates in space.

However, it should be kept in mind that when observing and knowing the inanimate world, a person (following the rest of the organisms) uses programmatic space-time and causal principles.

The modern picture of the world is physic-centric, where the laws of the non-living world are at the basis: ideas about the elements of matter formed by particles and fields, and their interaction. In this case, the main methodological approaches to the study and knowledge of the world order: the concepts of time, space and cause-effect relationships are a priori assumed by the inherent properties of the non-living world.

The living part of the world is supposed to come from its inanimate part, more complexly organized, but retaining all the patterns of the structure and behavior of this inanimate world.

The proposed concept of the program organization of living leads to a new look at the picture of the world.

The basis of it, as before, remains the basic laws of the non-living world – particles, fields and interactions between them. But the manifestation of these patterns in the living part of the world is fundamentally different in terms of the implementation of the interactions between the objects of this living world.

Instead of stochastic arbitrary interactions of equivalent objects of the non-living world, leading to random processes, in the living world interactions are ordered, regulated in time and space and are performed between nonequivalent objects, one of which is an active maker performing program actions in relation to another passive one.

A new quality of behavior of living objects is due to the presence in wildlife of an additional IS that determines both the spatial and temporal regulation of interactions and their causal relationship.

It is through this system that human, like all living organisms, learns about the world, including its inanimate part, and interacts with it.

Thus, in the proposed picture of the world, the main methodical temporal, spatial and cause-and-effect approaches to the study and knowledge of the world derived from the organization of its living part, namely its inherent IS. These approaches are not the result and manifestation of the laws of the non-living world. Moreover, since all the basic laws of the non-living world were studied using the "subjective" IS of the living, they bear the signs and imprints of this "subjectivity". This does not in any way question the authenticity of these regularities, although sometimes the clarification of their wording and interpretation is necessary.

Thereby, along with the general physical material structure – particles, fields and their interactions, the program structure of the living world and the IS providing it, come to the forefront of the basis world picture.

It should be noted that IS is an integral part of the program organization of the living. Therefore, to explore and formulate the features of the structure and functioning of IS, one should use features of program systems.

Already at the formation stage of independent volitional programming, in addition to clarifying and searching for new concrete effective areas of human activity, it was necessary to begin to build a more general system of relations with the outside world, including cult one.

At the time, a programmatic causal approach worked for the appointment of a responsible maker (idol) for the most important natural phenomena and the creation of the required situation for the execution of various VPs: rain or lack thereof, keeping fire, successful hunting or fishing, etc. Although this cause was not true, but only conditional, it was rigorously fulfilled, since it solved the important problem of strategic certainty and sustainability. This approach to the primacy of natural elements passed into the early religions, and then into the first philosophical teachings. No wonder fire and water excelled in the world picture of Greek philosophers.

With the development of scientific ideas about the outside world and its structure, more and more details appeared in the picture of the world, and the conditional causes became more meaningful.

Newton made a decisive transition to the modern physical basis of the picture of the world in the field of causality of separate bodies interaction and Gibbs formulated principles for describing the behavior of statistical ensembles composed of such bodies.

Regarding the conditionality of the principle of causality, applied to equally interacting bodies-objects of the non-living world, we have repeatedly expressed our opinion in the course of the previous presentation of the material of the book.

As an argument, examples were given of a purified, true scientific, and above all a physical experiment, in which one of the objects of research is "appointed" as active maker, and the other passive object on which the maker affects. In this case, the experimenter actually creates such pure conditions, when only one pre-determined active object acts on a previously selected passive one in the chosen place and at the chosen time. Thus, an ordered program situation is completely recreated in the study of a fundamentally stochastic equivalent interaction of non-living objects. The result is a very important practical and general scientific result that describes the pattern of such an ideal (essentially programmatic) interaction, or rather, a unidirectional action.

To obtain a more complete picture in the observed situations, when conducting a series of experiments, the active and passive objects of interaction are reversed, such as a magnet and a circuit with current, as in Faraday's experiments.

These comments do not in any way cancel the scientific and practical value of the results obtained in such purely programmatic conditions, but it seems to give them a certain conventionality, especially when studying the interaction of objects with stronger manifestations of stochasticity.

As for the statistical ensemble approach, it is based on the primary consideration of an ideal gas. A statistical set of identical, equally interacting objects-molecules is considered. And the assumption of "molecular chaos", i.e., stochastics resulting in uniform distribution of molecules in the space under consideration and their velocities and energies in all directions of this space. When describing the behavior of other ensemble, not only molecular objects, the main elements of the approach – the sameness of objects and the stochastic reproduction of their observed parameters remain. Ensembles consisting of several sub-ensembles are considered, but it is assumed that in each such sub-ensemble these two conditions are fulfilled.

At the same time, the approach allows one to consider the dynamics of development of stochastic processes, but only in the form of transitions from one equilibrium state to another.

As we see, it is impossible to construct such a statistical ensemble from obviously nonequivalent interacting elements of the living – makers and objects.

In addition, the essence of the information organization of programs from regulated actions is in direct and fundamental contradiction with the establishment of an equilibrium distribution of parameters in statistical ensembles.

These two fundamental circumstances, unfortunately, do not allow using the most powerful apparatus of statistical physics, which, in particular, is the basis for substantiating the three thermodynamic laws, for describing objects of living nature, at least within the limits of its existing concepts.

As for the presence in the stochastic world of inanimate relatively ordered structures and reproducible processes observed in natural conditions without a programmed research "staged" by a scientist, their very appearance is determined by the stochastic occurrence of a situation conducive to their origin. And after their fixation, it becomes possible to observe them, identify patterns and, in some cases, reproduce them in a program "formulation". For example, in the case of the formation of natural crystalline structures in local areas of a volcanic vent, spontaneously energetic conditions arise for the crystallization of a material localized in these areas.

Whereas stable for long observation times, the orbits of the planets were formed as a result of the action of stochastic processes of formation, for example, of the solar system.

All this is fundamentally different from the constant and regulated in space and time of a reproducible program living organism.

STRUCTURE AND PRINCIPLES OF VOLITIONAL PROGRAMMING – BASIC ELEMENTS AND OPERATIONS

Language and Volitional Programming

As noted in the first section of this chapter, there is an evolutionary synchronicity in the explicit manifestation of volitional self-programmed human activity, the emergence of Homo Programming and the appearance of linguistic Broca and Wernicke regions imprinted on the paleontological skull bones. Therefore, the statement that the formation of a language neural network was the basis for the emergence of a system of VPs does not seem to be groundless.

According to the author, the opposition of linguistic communicative function – the thinking one with the program approach ceases to be relevant. In this and in another case, the subject is external programming: compilation of programs from components – makers, their actions and objects, selection of sites and conditions for their implementation, as well as memorization and subsequent use of ready-made, developed programs and, if necessary, their modification.

That is, taking into account the material of the previous sections, the language is the basis of a person's personal volitional programming, the execution of collective programs and, naturally, all communication processes that permeate the functioning and development of an autonomous system of VPs.

Simply put, objects of human thought and communication are various aspects of volitional programming.

The sayings of the "great minds" who deny the role of linguistic symbols in their thinking process are well known. For example, Einstein formulated the following thought: "For me, there is no doubt that our thinking proceeds, basically, bypassing symbols (words) and, moreover, unconsciously". Thoughts about personal internal non-linguistic language when considering scientific, in this case specifically mathematical models and theories, expressed the founder of cybernetics Wiener.

It seems to us that in these cases, symbolic thinking is still present in a more general sense. But such symbols and their corresponding images can be not general linguistic, but professionally specific, for example, physical or mathematical, and then we can speak of branch languages of mathematical, physical, and even more narrowly, the language of quantum mechanics or statistical physics. And to the great scientists, they generally can be personalized, individual. But at the same time, the entire informational mechanism, based on the interactions of the pair of image-symbol, seems inherent in any manifestation of human volitional thinking.

Now it is appropriate to formulate the concept of thinking as a set of information operations and transformations with images and symbols designating them (of any linguistic and special, including individual, type) of all program elements and VP scenarios, as well as their signs and assessments necessary for reproduction, modification and drawing up new VPs, which constitutes all conscious human activity.

Human consciousness (self-awareness) is self-controlled work with VPs on their choice, reading, executing, modifying or composing new ones.

In connection with the above statement, in the future, speaking of language programs, we will focus on the functioning of analytical neural network blocks, rather than specific sensory auditory and effector speech ones.

At present, it is known that in addition to the main neural bundle connecting them with each other, numerous neurons from other brain regions also fit the language centers of Broca and Wernicke. This could be expected that the execution of language programs requires the creation and operation of an extensive and highly informative neural network.

In the past two decades, with the advent of magnetic resonance imaging (MRI) techniques, research began to be conducted on recording the activity of individual brain regions when a patient performs various behavioral and mental tasks of a researcher.

Thanks to these studies, the presence of a brain area in which the "zone of one's own will" was discovered is now established. This is the prefrontal cortex, part of the frontal lobes of the cerebral cortex, located predominantly under the frontal bones of the skull. Damage to this area often leads to characteristic behavioral disorders, including large-scale loss of the ability to make independent decisions.

In experiments using MRI, it was in this brain area that activity was observed in response to an independent volitional task of the subject, at his discretion, to perform one or another of his actions, movement.

In the presentation of further material, we assume that for the implementation of communicative activities, and in fact work with programs, common structural elements and operations with them are used, which we will consider below.

Images and Symbols

The basic element of VPs is an information pair (IP): an image and its corresponding signal mark – a word.

IPs forms a structural hierarchy. In one of the IPs may be included as components other IPs. And, on the contrary, the image with its own word can be part of another IP of a higher structural order. Both IP parts are information patterns consisting of information components. So, a word is a sequentially ordered pattern of sounds or letters, and an image is a pattern consisting of a set of features, details represented by other IPs of a lower level of the structural hierarchy and relationships with other images.

Typically, a word pattern causes, generates an image pattern, what, as can be assumed, is performed by a certain local network neural decider. The nerve impulses of the word pattern are fed to its inputs, and the pulses of the image pattern are generated at the output. Similarly, one informational pattern is converted into another one by cell protein p53 decider and by the multicellular analytic neuron decider. In all three cases, we can talk about the decider reaction on the input pattern by generation of the output one.

In the last two examples, the input signal pattern is a combination of factors and signs of a real, online internal and/or external situation, and the output is a set of reactive informational or effector actions.

In the case of VPs, the input pattern is the internal signal pattern of the word, read from the permanent library memory of a person at his will, will.

The output reaction is the appearance of a pattern-image corresponding to this word in a section of a neural network register that performs further operations with this image, for example, comparing it with other images or adding it in another one to create the image with a higher hierarchy structure.

This operational register (OR) should be able to online memorize for the duration of operations.

In the case of repetitions of such operations and at their high priority, it should be possible to memorize them in more long-term sections of memory.

It is significant that the contents of such an OR can, at any time, be read and analyzed by a volitional control signal for a subsequent decision reaction using additional network neural deciders.

Next, together with the "word", we will use the broader term "symbol".

An important property of a symbol-word (CW) is its ambiguity when used in different VPs, i.e., depending on the specific VP, the same CW reveals a different image.

We can talk about the block structure of the full image corresponding to a given word-symbol.

Each block has network connections with its own group of program elements, forming together with them local figurative networks determined by the type of a particular program.

To define an image block, a qualifying symbol is often used, usually consisting of several words. Thus, we can talk about the presence of composite phraseological symbols, phrases revealing the block structure of the generalizing image. Each word of the phrase corresponds to a specific image or indicates a network neighbor.

It is possible to form local image networks by signs, as well as by the types of actions performed by the image-makers.

A more detailed consideration of the structure and functioning of VPs should take into account the presence of a wide range of images in different classification systems. An example is static and dynamic images. The latter display a certain clip characteristic, the trajectory of the change of static images or a generalized way of actions and the courses of programs.

We can talk about real situational images, i.e., sensory, to varying degrees processed, and generalized, as they say, abstract images.

You can also note the presence of reactionary, effector images that follow, as we know, after situational ones.

Hierarchy of Personal Information Programs

Information programs include ones that manipulate all, without exception, VPs (including themselves), namely, retrieve them and their elements from the library, ensure their virtual execution, carry out their modification, evaluate the results of the modification and track their actual execution. Further we will use the term "program-manipulator" or simply the manipulator.

In addition to manipulative programs, information programs include the control, coherent, instructive, recording and reading information arrays and other programs already mentioned by us.

Further, it would be appropriate to consider the hierarchy of control programs in the VP system. The manipulation programs we have introduced just above can also be attributed to the control programs in that part of their competencies when they issue control signals for the execution of auxiliary information programs: reading program materials from the library, composing from them variable sequences and patterns, comparing options and developing a solution comparison results.

Based on the foregoing, the manipulating programs can be considered the highest in the hierarchy of the control programs considered so far, since the success of other information and effector programs depends on their adequate execution.

However, apparently, there is another higher level of control VPs. This is a volitional stimulating program (SP).

SP makes decisions about which program is to be launched and processed by the manipulating program.

For this, SP uses a system of program schedules, diaries, calendars, organizers that are an important part of the information libraries discussed earlier in the current chapter. These schedules have a temporary structure, displaying different time periods from seconds to months and even years. The most familiar to us are the daily program schedules, the so-called day regimen, which reflects the sequence of standard programs. Everyone is familiar with the lists of mandatory current nonstandard, attendant programs that need to be performed at certain hours and even minutes. Finally, there are strategic program planners of human life, marking its stages for the implementation of targeted training programs, the development of professions and the consistent achievement of increasingly high results and objects.

In addition, SP takes into account information about the conditions for managing personal programs, and their synchronization with other programs, both personal and collective, also contained in libraries.

So, the main task of SP is to choose the next VP launching.

At the same time, it has constant control over external and internal sensors that reflect the state of the body, as well as, if necessary, run emergency local programs, which are also recorded in the corresponding sections of the memory.

"In its spare time" SP is engaged in sorting and streamlining the AL, and, in particular, "stir up the memories".

All this aggregate information is displayed on the "internal display" of the corresponding section of the neural volitional network related to SP and is perceived by us as a "stream of consciousness".

In light of the above, we can talk about SP, as a program that generates consciousness.

Possible Schematic Scenarios of Stimulation and Manipulative Programs

After receiving a signal about the execution of the program, the next in the SP schedule, the manipulator program (MP) calls its image blocks from memory to the OR. The initial image information and subsequent manipulations with it are observed on the neural "display, monitor" in both MP and SP programs.

Apparently, the script of the next program with the basic, for example, the elements most frequently used earlier, primarily makers and objects, turns out to be primarily on OR. The data characterizing the program and stored in memory with it are compared with the strategic and operational conditions of execution coming from SP. If there is no conflict between them, the next program is executed in the basic version.

If such a conflict is detected, for example, the execution time of the next program in the basic version is too long, which leads to interference for other actual programs, the MP performs manipulations with the basic regular program, reducing its duration in different ways. This may be a reduction in program steps or even the removal of one of the constituent programs. The next program modified in this way is analyzed again for conflict issues with other programs and, in the absence of such, is performed in this modified version.

Unfortunately, the currently available information is not enough to formulate any reasonable options for specific scenarios of local programs included in the MP and SP. Therefore, we confine ourselves here to only a brief list of functional operations, the performance of which seems necessary to solve the above tasks of these higher control program hierarchies.

- Scanning the program schedule and selecting the next program to perform
- Reading the selected program to the OR
- Manipulations with the OR program leading to its modification, for example:
  - Permutation of actions and components of programs in the scenario, their reduction or addition
  - Replacing makers and objects
  - Site replacements

- Evaluation of the characteristics and parameters of the program and its elements, taking into account the online conditions for its implementation, for example:
  - Execution time
  - Availability of makers and objects
  - Site availability
  - Energy sufficiency
- View and virtual run of the selected and modified programs
- Launching a program version adequate to the conditions for its implementation

All of the above operations require the presence of those additional volitional network neural structures that allow a human to individually determine the sequence, type and modification of a regular external program or create a new one.

At the same time, it is logical to assume that the same information principles work in these local volitional networks, which ensure the execution of external programs for multicellular ones, namely, processing and recognition of image patterns with the subsequent decision to connect informational or effector makers.

For the work of volitional control programs, the most important is the possibility of prompt volitional change in the setting of a neural decider, both at the input and at the output.

It can be assumed that this possibility is provided by the plasticity of synapses, controlled by signals from stimulating and manipulating programs.

INHERITANCE AND DEVELOPMENT OF UNICELLULAR AND MULTICELLULAR INFORMATION SYSTEMS IN VOLITIONAL PROGRAMS

Considering the information component of VPs, we are finally convinced of the unity of the basic information principles for all evolutionary stages: unicellular, multicellular and human. How do we joke in such cases – and what did you expect? Indeed, the last two stages are essentially the evolution of the organization of cell communities, and what informational principles can develop in them, beside those which work in the evolutionary "instigator" itself – the cell.

Let us once again go through these principles.

First, we note such a general property of program systems at all evolutionary stages as their network organization. There are local program networks organized by tasks and results.

Here we should clarify the concept of "network of programs" in comparison with our earlier term "program system". In the first case, the presence of interconnections between programs that determine their interdependence is implied. It is these links that lead to the synchronization of the execution of different programs in time and intensity, which ensures a single, common result for them. But this relationship does not imply the use of the term "system of programs".

Therefore, the use of the term "network" instead of "system" is not only new, but also more correct and productive.

Network interconnections between programs provide all the same informational signal marks that may be the result of special information programs, specifically the work of intelligent makers-deciders, or the product of a neighboring, spatially or thematically, program, as is often the case for cellular programs.

The transition from unicellular to multicellular and further to human increases the complexity and efficiency of networked program structures and at the same time their role in the creation of neural networks, an integral part of the overall programmatic network of the organism.

It seems to us that the division of the information structure into the instructive reproducing and signaling, controlling, adaptive structure introduced in Chapter 1, has justified itself.

Reproduction of programs at all three evolutionary stages is mainly determined by gene instructions, to which a person's volitional "man-made" program instructions are added.

From our point of view, the second signal part of the information structure, which integrates the body's programs into a harmoniously functioning program network that adequately responds to changes in internal and external conditions, is of particular interest.

The main basic element of such an organized program network is an intelligent information maker – decider. In the cell, the most striking example of a decider is the p53 protein, and in MOs and in humans, a neural decider, the role of which is played by a section of an analytical neural network or, in the simplest case, one, the so-called inter or command neuron.

In all the considered examples, deciders have many inputs and outputs that allow them to respond to the input situational signal pattern with a reactive output pattern, which can be informational for subsequent analysis or to run effector programs.

So, a pattern consisting of a set of situational signals, if it coincides with the setting of the decider pattern receptor, causes an internal recognition signal, which can be interpreted as a convolution operation of the input multi-signal, often multimodal pattern into one internal trigger signal, the excitation potential in the neural decider. Further, this signal diverges, turns around at different outputs of the decider, activating various information and/or effector makers, thereby performing the output pattern sweep operation in accordance with the configuration of this decider. A pair of input and output information patterns, solving a specific situation in the only adequate way, characterizes each decider.

It is known that it is possible to reconfigure the decider as by the input as output. In the p53 protein, this reconfiguration can be caused by special signals that change its covalent modification. For neural deciders, the mechanism of the so-called plasticity of synapses presumably works, as a result of which the input and output patterns of neurons and local area networks consisting of them can be modified.

The need for different options for responding to different situations is obvious, and the ability to use one decider in such cases, adjusting it to new situations, is a much more economical way than having a decider for each situation. Although for the dynamic virtual network of neuro deciders, this last "straight-line" option does not seem hopeless.

On these briefly considered general basic information principles, evolution has created in humans a superstructure of VPs. Why superstructure? Because the entire IS of higher multicellular, for example, primates, exists and works with minimal changes in humans in the subsystem of internal programs and in the "intuitive" part of external programs, manifested, for example, in the hero "Mowgli" or in the "automatic" return home of the drunk man.

Based on the material in the previous section, we very briefly recall the main informational "innovations" of VPs.

First, is the emergence of a secondary internal signaling system, essentially a language one, the basis of which consists of pairs: word-image of a program component in the broadest interpretation of this image. Such a wide interpretation of the image – "everything about the program" includes tasks and results, scenario, scheme of actions and connection of auxiliary and additional programs, makers, objects, program execution sites, energy sources, etc.

At the same time, the image presumably has a block structure, including belonging to a particular program characteristics, properties, parameters and ratings, etc.

This information is placed in memory in such a way that it forms associative hierarchical networks, allowing you to produce when reading a grouping of images according to the contents of different image blocks, for example, choosing red fruits, the fastest cars or makers for processing of wooden details.

All this reminds us of modern Internet search engines. You should not be surprised, since they are made in the image and likeness for the brain network of VPs.

In addition to the library of the secondary signaling system, a significant novelty of the human IS is new programs that implement work with this library, primarily reading and writing information, as well as manipulating symbol-image pairs for evaluating and modifying programs and general managing of VPs.

Apparently, these programs and the corresponding program networks are implemented in the brain areas of Wernicke and Broca, as well as in the "zone of one's own will" revealed in MRI studies. A symbolic-image library is apparently also located there.

It can be assumed that the basis of the structure and functioning of these additional parts of the human brain are neural network formations similar to those that in MOs provide transformation of information patterns.

So far, we have been talking about the evolutionary brain part of the informational organization of individual, personal VPs.

However, already in the very early period of the existence of Homo Programming, the formation of collective information programs was to begin, helping to work with the program library, modify and create new programs.

Such informational findings were collectively tested and remembered in the socialized memory of the community. The basis of such a first collective IS was spoken language.

The emergence of the written language served as a huge leap forward in this area, since a reliable base appeared both for creating a collective program library and for editing and creating new VPs.

Nowadays, the role of collective information programs has grown immeasurably.

The directions of development of information technologies are the same as tens of thousands of years ago at the evolution beginning of the VPs. These are memory devices, means of transforming information patterns and arrays, means of exchanging information data between information subjects.

We can say that the program-information approach turned out to be effective in considering all three stages of evolution. The evolving maker structure performs more and more complex evolutionary promising programs. An information structure that provides reproduction, interconnection and synchronization of programs and performs a very important function of developing adequate program responses to changes in internal and external situations is becoming more efficient and flexible and very importantly much faster.

An IS is being developed on the same base, consisting of pairs: an information mark – a recognition maker domain and operations carried out with these IPs.

Naturally, this information base, embedded in the cell, first manifested itself in the formation of the structure and principles of functioning of the brain department of human VPs, and then transformed into the main directions of collective information VPs.

It seems logical to attribute to the system of information VPs not only those scientific and technical fields that are directly involved in the development of tools and algorithms for memorizing, processing and transmitting information, but also its scientific basis, primarily mathematics.

It can be assumed with a high degree of confidence that, elaborating methods and methods of working with VPs, a person has developed criteria for evaluating their parameters, properties and characteristics, in particular, based on mathematical systems of measures and sets. The basic principles of such criteria and evaluations work in a cell, because without them, neither reproduction of programs nor their control can develop adequate responses to changing situations.

As for the "physical" informational principles of space, time and causality, then, again, their use, especially in external multicellular programs, seems necessary. And the fact that they acquired special significance in strong-willed external programs and have great scientific interest is not surprising.

Thus, in conclusion, I would like to confirm here the thesis from the end of Chapter 1.

The information system of all living things includes information programs and representations that ensure the reproduction and functioning of all body programs.

4 Program-Information Unity of Living

COMMON PROGRAM AND INFORMATION STRUCTURES

The commonality of all considered evolutionary stages: unicellular, multicellular and human is their program organization as follows from the previous three chapters of this book.

The author's acquaintance with plant representatives of nature, not reflected in the book, allows to make a more general conclusion about the universality of the program approach in relation to all living organisms without exception.

The basis of this approach is the statement proved by the materials of the book that all actions taking place within a living organism or committed by itself in the external environment are the result of a predetermined, orderly transformation of the passive object properties by an active participant of an interaction – the maker or its agent.

The set of such actions, leading to the result: the division of the object, the assembly of the object from the component parts, moving the object is called the program.

Apart from the maker and the object, the components of the program are the site – the place of the program execution, and the energy necessary for the maker to perform an action on the object.

The final task results are highlighted on the background of many intermediate program results. They are practically identical in all species of organisms and consist of the following main groups: manufacture of makers, getting and storage of objects and energy, formation of sites.

Obtaining task results is carried out usually by composite programs that include more simple programs or individual actions.

Complex of the task programs should contribute to the achievement of the main goal – to ensure its reproduction and functioning in a changing environment.

To do this, the reproduction program, as well as a group of "survival" programs that include: the maintenance of homeostasis inside organisms, i.e., the stability of the basic conditions necessary for the implementation of all programs; countering the accumulation of stochastic defects in the organism; the fight against parasites and other biological enemies of the cell are added to the above-mentioned task programs.

In addition, important programs are those that provide adaptability of organisms to changes and adverse effects of the environment both in the tactical time range within the life of the individual organism, and for a long time of change of many species generations in the work of the evolutionary mechanism.

Signal and control programs provide a coordinated, synchronized operation of the program system of the organism, necessary for the timely passage of its development phases and adequate response to changes in internal and external situation. They are included in information program subsystem of the organism.

Information and effector program subsystems form a complete system of organism programs.

Information programs provide control, synchronization, cooperation of all cellular programs, producing, transforming and using information in the form of signals and instructions.

The basis of the information system is a pair of mark and recognizing maker domain. When a mark gets into the corresponding domain, their information interaction occurs. Its result is the change of maker state from activated to inactivated one (or vice versa) or emitting its signal information mark.

Information marks form patterns of gene DNA and RNA instructions for the manufacture of intracellular makers. The signal marks, forming their pattern structure, control the frequency of programs execution, the most important characteristic of the organism program system. Identification marks allow the maker to recognize its object.

Essential elements of information system are information makers – deciders that in the information signal chains take place between sources of signals, receptors and endpoints of their which launch target effector maker. A decider, recognizing the input signal, decides to run or not to run the effector of the chain.

Often a decider having multi-signal input and output reacts of its output signal or effector pattern to situational input signal pattern.

As a result, a network of interacting effector and information programs is created from the system of individual organizational programs that solve the regular tasks of the organism and respond to changes in the internal and external situation in a coordinated and effective way.

Thus, each organism can be characterized by its program system or network.

The classical approach uses the concept of genotype and phenotype to characterize the organism.

Genotype – a set of genes of the organism.

Phenotype – a set of all the individual organism features, formed in the process of its genotype and the environment interaction.

In relation to the approach presented, you can introduce a third notion of "programtype", abbreviated "protype" – organismic system of programs that are implemented throughout the life of the organism.

From the standpoint of the program approach during the earth's evolution of life, there were two fundamental leaps in the change of the organism protype – the emergence of multicellulars and the appearance of man.

Both leaps are essentially due to the "innovations" of the cellular organization.

THE ESSENCE OF CELLULAR EVOLUTION

The most significant evolutionary changes of the cell, which led to the transition from prokaryotes to eukaryotes, did not affect the essence and composition of the program system. Although there was a significant shift, for example, in the energy supply when mitochondria have been included in the cell composition, on the necessary and sufficient program set and their organizational principles it didn't affect qualitatively.

The real evolutionary "revolution" began when the cells, communicating with each other on common species issues, "realized" that they live well, when they are temporarily combined, forming so-called aggregate organisms, especially in times of famine.

The further development of such intermediate organisms led to the creation of social cellular communities in the form of a fundamentally new type of multicellular organism (MO).

To implement an effective social community of cells, two crucial problems should have been solved. First, it was necessary to create a mechanism of cell differentiation to perform programs for different purposes of MO. Secondly, it was necessary to provide a unified social system of exchange of the cellular program results for the total guaranteed supply of material resources and oxygen, as well as signal cell communication within the whole MO.

MO can be considered certainly as qualitatively new kind of protype compared to unicellular, formed by cells, with the former, though modified, program organization.

As for the next stage of evolution – human, there are relatively small, local evolutionary changes in the brain of the same MO led to a qualitative, radical change in its program capabilities. For the first time, there was a biological species that can independently, by oneself, program all its behavior. As a result, the speed of evolutionary adaptability has increased millions of times.

It should be noted here that the man is the same cellular social community as other MOs.

At the same time, human individuals unite in their social communities of the next hierarchical level, in which the effectiveness of independent, volitional programming increases many times.

Let us now dwell on all three stages of evolution.

CELL AND UNICELLULAR ORGANISMS

The traditional definition: "cell, the basic structural and functional unit of all living organisms" in the program approach to life acquires an additional meaning. The fundamental principles of the "living" program-information structure are laid in the cell.

Let us briefly recall this structure.

MAKERS

Proteins, ribonucleic acids and high-molecular complexes based on them perform the role of makers in the cell.

The main types of the maker's actions are mechanical: object separation, object assembly and object movement and chemical, leading to changes in its chemical state, for example, by phosphorylation, acetylation, methylation, etc. It should be noted that most of the actions are complex – chemical-mechanical.

The maker can be "transforming" performing the transformation of the object state, and "structural" forming the structure of another complex maker (e.g., ribosomes or microtubules) or membrane and cytoskeleton.

In some cases, in addition to the makers, the programs can be executed by their agents – ions, atoms and molecules generated and/or transported to the site by the respective makers. Such agents include, for example, calcium, potassium, sodium, etc., ions.

Usually the maker has several functional domains.

It is primarily an "object domain" that captures "his" object for later performing a program action on it.

In addition, there are several types of information domains.

These include areas of makers, in which information agents – molecules or ions (an example of which is traditionally considered coenzyme) activate or inactivate this maker.

There can also be domains that recognize the characteristic of "his" object, for example, the promoter of the gene.

A typical behavior algorithm of a simple maker includes: the activation of maker by an information agent, trial contact with an object, object recognition and the formation of a working contact between a maker and an object, the implementation of the program action of the maker over the object, the separation of the maker and the object modified as a result of the action. The energy necessary for performing a program action comes to the program site from the surrounding cytosol or is brought with the object.

Program Structure

The object modified according to the type of action is the result of the program. As a rule, the result of the program action performed by one maker is the object for the next maker's program.

Programs executed by one maker are called simple programs.

In the case of a simple maker, such programs are one-action or one-step ones. An example of such programs is separate catabolism program of the Krebs cycle.

If a simple one-maker program is executed by a complex maker, then such program is multi-step, for example, a program executed by a ribosome.

A set of two or more simple programs that lead to a single result and committed in one site, we call composite programs.

Composite programs, according to the definition of a simple program, are made by more than one maker.

Examples of composite programs can serve as the Krebs cycle, consisting of a simple one-step catabolic programs, and assembling a peptide chain operated by the complex maker – ribosome and simple maker – transport RNA.

The result of a composite program is a task result in some cases.

However, in many cases, the task requires running multiple composite programs in different sites. For example, the task of obtaining a protein maker is solved by performing two important composite programs – transcription in the nucleus and translation in the cytosol of the cell.

The set of programs that lead to the task result, we call the task program, an example of which can serve as considered in Chapter 1, sequentially executed programs of transcription, RNA processing, translation, folding peptide chains and, finally, assembling of a ribosome from RNA and proteins. The result of this task program is the production of a complex maker – ribosome.

The most important criterion, characteristic of the program is its effectiveness, i.e., the number of implementations of its result per unit of time. It is the relative partial effectiveness of a single programs in the cell that forms the state of the cellular program system, which, as we know, determines the functioning of the cell organism.

Therefore, the key direction of management of individual cell programs is the regulation of their effectiveness.

Management of Programs

The effectiveness is determined by two parameters: the pace, speed of a single program and the frequency of its playback.

Usually, the standard speed of program execution varies within small limits.

In this regard, the main control mechanisms are focused on adjusting the frequency of the program.

The frequency of program reproduction is determined primarily by the concentrations of the makers, their objects and energy particles on the program site.

The management of enzyme maker concentration is the determining and most often considered in the literature factor.

Two main programs – producing makers and transporting ready-made makers on their "working" site provide the required maker concentration on the site.

The most important of them is the first producing program and in it's a composite program of transcription.

It is not surprising that there are a lot of mechanisms for regulating the frequency of this program in the cell using combinations of different regulatory proteins interacting with the promoter part of the transcribed gene.

Frequency control of various cellular programs – one of the most widespread, but not the only local task of control programs.

Information Network

The common task of such programs is to harmonize all cellular programs for joint response to internal and external situational signals.

To solve this problem, there are signal chains starting with the receptors of the primary situational signals, ending with information makers launching target programs, and having intelligent makers deciders in the middle between them. Such chains form information networks. The most efficient way to create networks is facilitated by the presence of deciders with multi-signal input and output, an outstanding example of which is p53 protein. An important feature of this protein is the ability to change its functionality, due to covalent modifications caused by various protein deciders controlling it, that stimulates the formation of information networks in the cell in turn.

Tasks and Structure of Information Programs

Among the main tasks solved by information programs (IPs) in the cell, we can distinguish the following:

- Playback of makers specialized for a specific object
- Identification of objects and their delivery addresses
- Optimization of individual program execution frequency

The key action of the IP is recognition of "own" mark by the maker domain and the subsequent reaction of this maker in the form of its activation/inactivation, or the direct implementation of its inherent action.

There are three types of marks: instructive, identification and signal.

Instructive marks of DNA and RNA contain information on which components of all makers (proteins and RNA) are assembled.

Identification marks are mainly used for object identification by the maker.

Signal marks form program signal chains combining in the network to solve the problem of program synchronization for proper response of cells to changing external and internal circumstances.

Consider some actions and programs in which marks of different types take a passive participation (as objects) or active (as makers).

Among the programs performed by the makers in relation to passive instructive marks, it should be noted:

- Replication – copying DNA instructions to daughter cell
  - Transcription – conversion of genetic information from DNA code to RNA code
  - Repair of damaged DNA instruction elements
- Recombination, the introduction of evolutionary variation in DNA instruction
  - Finally, the central program is the translation – reading of RNA instructions and production of peptide chains, semi-products of the most common makers: proteins and complexes based on them.

In view of their results, we will refer them all to the effector programs.

Important programs of the promoter DNA region modification using regulatory proteins that largely determine online status of cell program systems should also be entered in this group.

Simultaneously with the implementation of the first two programs, instructive marks reproduction occurs.

The object identifier mark performs a passive role in relation to the maker also because it does not change the state of the maker, but only contributes to the performance of their inherent action on the object.

In contrast to the first two types of marks, the signal mark changes the state of the maker and can therefore be considered as an active element of interaction with the maker, which in this interaction plays a passive role of the object. As we pointed out, the role of the maker, i.e., an active participant of the program action, in some cases, can be played not by makers themselves, but by agents, which are products of other makers often.

Simple marks form composite information patterns, which in turn can form more complex patterns of higher orders of hierarchy, as observed, for example, in gene instructions and situational signal patterns.

Possible operations with the information marks include the following:

- Convolution of the pattern into a mark or a simpler pattern
  - Deploy of the pattern
- Recognition of a mark and a pattern with subsequent reaction of maker
- Storage of a mark and a pattern

In addition, the information network formed by deciders can perform logical operations with signal marks: negation, conjunction, disjunction, and to solve with them quite complex logical problems for making reactionary decisions that are appropriate to the current situation.

The following is a definition of an information system:

Thus, the information system (IS), which includes IPs and information actions in effector programs, is used to reproduce, interconnect and synchronize programs, as well as to work out phases of the cell cycle and adequate reactions to changes in internal and external situations.

Base of IS is information marks and maker domains that recognize the marks.

The essence of IS – operations and programs performed with the participation of IS, which can be represented in terms of concepts and representations of traditional informatics.

The space-time concept and the principle of causality are also included in the IS, which permeate the entire regulated program cell organization.

The openness of the IS definition is the generalization of all means of reproducibility, regulation and adequate response of programs.

Complete System of Unicellular Programs

The purpose of the program system (PS) is the implementation of many phenotype self-reproduction cycles to provide its adaptive changes, including through evolutionary changes in the genotype.

To achieve these goals, PS should solve the following tasks:

- Maintain the number of groups and programs in them at the required and sufficient level. At any given time to be able to reproduce and vary the intensity for any program of the system, providing the necessary level of material and energy resources
- Maintain the conditions necessary for the operation of the programs, primarily by providing cell compartmentalization and implementing homeostasis in these compartments
- Provide the necessary rate of cell reproduction cycles

- Provide anti-entropic updating, purification, correction and treatment of program structural elements, especially makers and compartments
- To adjust PS and phenotype to changing conditions
- To implement control of PS programs, to maintain regular, adequate to this phenotype, status

The achievement of the main goals of the organism is implemented by the following groups of programs, united by their task result.

1. **Maker programs** that produce makers of all programs: proteins and RNA, as well as complexes of them as homogeneous, for example, microtubules formed by self-assembly from one protein, and heterogeneous, for example, ribosomes collected from dozens of different proteins and RNA.
2. **Resource provision** programs provide the necessary material (molecular and atomic) and energy (in the form of energy molecules and electrochemical potential) base for the implementation of all programs. This group includes such external programs as the endocytosis and chemotaxis in his attractant part.
3. **Transport programs** provide transfer of makers, objects and energy molecules to their sites, as well as the work of signal chains and energy pumps.
4. **Compartment programs** form all cell membranes, including the outer and organ membranes, as well as elements of a cytoskeleton.
5. **Homeostasis programs** monitor the stability of the basic conditions necessary for the implementation of all programs: temperature, osmotic pressure, acidity level, etc., in the compartments of the cell.
6. **Signal programs** carry out the detection of signals that are specific molecules or physical effects, such as radiation, temperature changes, etc., as well as signal conversion, amplification and transmission to effectors. Signals characterize the state of both the external environment and internal parameters and characteristics of the cell.
7. **Management, control programs** carry out changes in the intensity of individual programs in accordance with the situation, determined by a set of both external and internal signals. For the most part, the control programs modulate the concentrations of the makers: proteins, RNA and their complexes in the active state. Such modulation is often done either by activating/inactivating the maker or by regulating his produce programs at various stages, primarily transcription and translation.
8. DNA replication and realization of cell division, mitosis include into the group of **reproduction programs**.
9. **Anti-entropic programs** (AEPs) counteract accumulation of entropy of the phenotype and genotype of an organism due to stochastic degradation processes and ageing. This is primarily a program of the ubiquitin-dependent protein proteolysis. Further, AEP includes lysosomal and exocytosis programs for the removal of toxins and waste from the cell. Of course, the most important AEPs are reparative programs that restore degrading DNA information.

10. **Protective programs** include, first, immune programs to fight parasites, especially viruses. It is also a repellent chemotaxis program, heat shock control and other programs that protect single-cell organism from adverse effects of the environment and biological enemies.
11. **Evolutionary programs**, such as, for example, variable splicing of primary RNA and DNA recombination.

The considered groups of programs constitute *a necessary and enough program set*, allowing a single-celled organism to solve all the problems to achieve the above goals.

Cellular Evolution

Above we have given a complete set of programs necessary and enough for the evolving existence of cells for at least the entire observed time of evolution, i.e., several billion years.

The absence of one program group in the system at least should lead to irreversible destruction of the entire system of cellular programs and the death of the organism.

That is why, going along the evolutionary chain to its beginning, we rest against the earliest prokaryotic organisms that already have the entire set of programs we have considered.

It becomes evident that all intermediate forms between inanimate matter and the first ancestral cell with a complete program set are not capable of sustainable multiple reproduction required for the evolutionary process.

In particular, it is impossible to have a cell without a membrane. The expression "at a certain stage of evolution, the cell had got a membrane" makes no more sense than, for example, the following: "at a certain stage of evolution, the frog had got a skin".

To the well-known expression of Frazzetta in its "Complex adaptations in evolving populations" that the evolutionary problem is, in fact, how gradually improving the machine without interrupting its operation, it should be added that the machine must operate from the very first step of evolution, for what it must have from the very beginning a necessary and sufficient set of programs discussed above.

Here it should be noted that the machine is improved from the inside, by "germination" and fixing adequate, profitable changes in the PS.

Finally, the impossibility of the stochastic origin of the living from the non-living follows from the more general statement that the process cannot spontaneously transform into a program, because, as we have seen any program can exist only within a PS consisting of the necessary and sufficient set of individual programs.

Evolutionary leap did not come of cell structure improve, but how the implementation of the cell social integration into MOs.

MULTICELLULAR ORGANISMS

Structure of Multicellular Organisms

MO is a social association of specialized cells that perform their specific programs to provide general task programs of MO. For example, the cells of the stomach performing its social function by synthesizing and secreting of the components of gastric juice are involved in a common task nutrition program, vital for all cells in the MO without exception.

Among the tasks solved by MO, it is possible to distinguish the maintenance of life of all MO cells, providing them with the necessary components of nutrition and oxygen, as well as the implementation of immune protection from pathogens.

Cells, spending their resources to implement their specific social programs, receive gain in the form of guaranteed supply of resources and immune protection. Thus, the social behavior of the cells in MO makes it resource-efficient and evolutionarily promising, compared with unicellular ones.

Social cellular programs perform specialized social makers.

The cells containing social makers are called differentiated, and the programs of social maker formation in them are called differentiating. Thus, the cell in the MO has its own specific social makers that perform, in fact, external programs in relation to this cell itself, but internal within the whole MO. "Standard" makers performing internal cell programs, often referred as "housekeeping" in the literature, have much common for all types of organism cells.

To perform the general task programs of MO, its cells form a multi-level system of social makers.

It includes the molecular "elementary" makers (EMs), makers – cells or cell makers (CMs), makers – tissue or tissue makers (TMs), makers – organs or organ makers (OMs), makers – system or system makers (SMs) and, finally, integrated maker association (IMA) to perform the full task programs, such as nutrition programs. TM, OM and SM are composed of EM and CM, and therefore we will call them composite.

As examples of composite hierarchical makers can be called muscle and glandular makers. The first consists of synchronously working muscle cell fibers, which in turn contain elementary protein makers: myosin and actin, initially causing muscle contraction. Gland organ consists of secretory cells that produce a secret maker and then exocytosis program displays the secret in the intercellular space.

Even though in all the makers of the highest hierarchical level "real" work on the program implementation is performed by EMs ultimately, the above structure is not formal. It is essential that in many cases EM and CM execute programs as part of higher order makers synchronously with each other and in coordination with those included in other component makers.

At the same time, structural makers of CM and OM perform important functions for lower hierarchy makers: CM for EM and OM for CM. They include sites for the implementation of the main programs of lower makers, and also contain auxiliary makers, which primarily contribute to reproduction, proliferation and/or maturation, formation of the main makers.

Tasks of MO Programs

The basic task programs performed by somatic cells individually or as part of MO organs and systems are, first, supply programs of nutrition and respiration, reproductive program, protective immune program, as well as sanitary excretory programs of kidneys and large intestine. The most important social programs for the exchange of material resources and information signals between the makers of all levels are carried out by the circulatory and nervous systems of a MO.

Here are two examples of the most important MO task programs and their performing systems.

The task program of cellular nutrition includes three components.

This is primarily a program of getting food and its pre-fragmentation, which is performed outside the MO, thus being external, and is implemented by external bodies – makers, which primarily include the organs of motion, capture and pre-fragmentation of the food object, as well as sensory organs.

The following one is a composite program that is already internal, placed within a MO. In carrying out this program, its OMs, included in the digestive system of the body, primarily the stomach and intestines, carry out the grinding and partial dissolution of food fragments and, finally, bringing them to the level of components consumed by cells, i.e., proteins, amino acids, fatty acids, trace elements, etc.

The third final program, being transport, delivers the nutrients consumed by cells from the digestive system to the intercellular space, where cells absorb them with the help of their external endocytosis program. The main makers of the circulatory system – the heart and blood vessels, perform this transport program.

The main program of respiration provides the main component – oxygen of cellular energy programs to create high-energy molecules and ion gradients because of oxidation of nutrients. This program is performed by two complex SMs – respiratory and circulatory systems of the body.

The same organism systems perform an auxiliary respiratory program to remove from the body the final product of intracellular energy programs – carbon dioxide.

Management of MO Programs

Control of makers on all levels is possible provided that the control signal reaches the EM inside the cells.

Most of the MO cells are surrounded by intercellular space filled with interstitial fluid, which is the external environment for them. The exchange of substances between blood plasma and interstitial fluid occurs through the extensive surface of the capillary walls. Synapses of effector neurons secrete neurotransmitters to control muscle, secretory and other "target" cells into interstitial fluid. Thus, from the intercellular space cells receive "material supply" in the form of nutrient molecules and oxygen, and signal marks to control their programs from the nervous and humoral systems.

However, into the same interstitial fluid cells also emit the products of their elementary social makers and signal marks for the nervous and humoral systems.

Here it is important to emphasize that all MO cells are included in the information exchange by signal marks through the intercellular space and further through branched nerve and humoral (blood) networks. They perceive control signals through their surface receptors facing the intercellular space and secrete the control signals for the other cells and the information signals of their condition into interstitial fluid for later decisions of control neurons.

Nervous System

The main control system in the MO is the nervous system.

Being a social formation of specialized cells, MO inherits cellular information principles. The similar structure of networks formation from signal information chains is reproduced. The chains include three types of links: receptors, generating primary signals; analytical ones processing signals and making decisions in the form of reactive signals; effector ones sending the signals on target effector makers.

The main difference between the information networks of MO and cells is the structure of chain links. If there are elementary protein makers in the cell, then CMs – neurons appear in MO. An important function of the neuron is a fast signal transmission over a long distance comparable to body size. For this purpose, an electric propagation mechanism of the membrane potential pulse along the neural cell is used.

At the same time, signal marks at the input and output of the neuron have the same ion-molecular character as in the cell. In the input information domain (input synapse) of a neuron, such a chemical signal is converted into a pulse of the membrane potential that propagates at a high rate to the output neuron synapse, where it is converted into a chemical form again.

Like in the situation inside the cells, the neural information chain forms information networks with the crucial role of inserted control neurons, which we will call the intellectual neurons or n-deciders, to distinguish them from cell protein deciders.

Depending on the configuration and intensity of the input signals coming to their dendrites, n-deciders produce or do not produce an output signal on their axon triggering an effector maker or the next information maker in the chain. In a more complex cases, n-decider able to respond on known (recognizable) input signals by the configuration output signals to different locations.

Due to the possibility of implementation, two polar effects on neuron – excitation and inhibition, all the basic logical operations with signals: conjunction, disjunction, inversion, the implication can be performed by n-decider.

In addition, analog arithmetic operations of summation and subtraction can be used to estimate the resulting effect on a neuron.

In accordance with the modern view, a nervous functional structure consists of local neural networks formed in the peripheral and central parts of it, in accordance with the tasks that they solve – from the simplest reflex to complex behavioral one. The compound networks performing complex behavioral programs consist of a variety local sensory, effector and analytical networks of different complexity and hierarchy.

Network structures can be divided into virtual, temporary, formed at the time of solving given task, mainly in the performance of external programs, and anatomical, constantly existing in the body clusters of neurons. These latter include nerve centers, ganglia and nerve nodes, the spinal cord and brain, each with its own network structure. Such innate networks perform two main functions. First, they directly autonomously solve many so-called reflex problems of the lower level, for example, ganglia and nerve nodes in the digestive tract control its peristaltic and suction programs. And, secondly, they provide the original basic set of neural clusters and the main links within and between them for the subsequent creation of actual virtual network configurations.

Two mechanisms used for the formation of network structures.

The first of these, convergence is the integration and processing by a single neuron a plurality of input signals arriving on its dendrites, for example, from individual receptors or their groups. On axon of such n-decider, resulting signal occurs in the case of matching input configuration with preliminary stored one. Then the signal is sent to intellectual neurons or effector makers directly.

A second network mechanism, divergence, may also be involved. In this case, the signal for a variety of branches (collaterals) of the axon is sent simultaneously to the inputs of various "consumers".

Features of the Information System

Apparently, the greatest development of the network IS of multicellular, especially in higher representatives of mammals, was due to the growing diversity and evolutionary significance of external programs that are very limited in unicellular.

The set of individual receptor makers has been transformed into sensory OMs: visual, auditory, olfactory and other, issuing multi-bit signal patterns about the situation on the sites of external programs.

Accordingly, the analytical intelligent local networks developed from n-deciders transform and analyze the signals of sensory information and control the effector muscle makers.

Often, to illustrate the joint work of the sensor and analytical blocks of the information network, an example of the visual channel is used.

The recording image falls on the first receptor layer of the eye retina, where it is converted into a set of nerve impulses at the receptor synapses.

The first stage of image processing is the selection of the main information elements: dark and light spots and contrast strips located at different angles, carried out by neural CMs of the retina and primary visual cortex.

The second stage – the compilation from the selected image elements, the whole information pattern, which is often the outline of the object, is carried out by neural CMs of the secondary visual cortex.

The third stage of the analyzer maker work – comparison of the current information pattern with the ones memorized in the previous program implementations is carried out in the same place probably.

A variant of recognition can be the operation of one intelligent neuron, launching local task program, when entering on the synapses, the same signal components of the current pattern to which they were configured by previous successful program implementations.

At the same time, signals from the components of the current pattern can arrive simultaneously on several intelligent neurons that are configured to different remembered patterns corresponding to them and launch various subsequent programs. The implementation of current pattern recognition with the launch of a subsequent program takes place on the neuron, the setting of which corresponds to this pattern. In this case, we can say that the result of execution of the image recognition program of the current situation is the launch of the next local program, specific and adequate for this situation.

Memorizing and Learning

Each control task of MO has its own program that runs by the compound neural network maker. The current online state of a MO corresponds to a set of local neural networks that are changed and modified, adapting to changing conditions and specific phases of the organism programs. In this case, each network has its own time of fixed existence (TFE), providing dynamic stability of the body.

TFE corresponds to the traditional concepts of memory and memorization, used in the consideration of the nervous system functioning.

The range of TFE or using the generally accepted terms of storage time of modified neural network state is very wide – from seconds to tens of years. Short-lived networks perform operational control programs that provide solutions to online tasks. The longest lived networks provide strategic age stages of the general organism program.

Currently, various mechanisms of neural networks modification, because of changes in the neuron response to the signal, are actively investigated.

For example, it is known that the intensity of the neuron response gradually decreases with prolonged exposure to the signal. This phenomenon, called adaptation, is caused by the action of potential-dependent calcium channels and Ca^{2+}-dependent potassium channels.

When exposed to a sharp stimulus, such as an electric shock, the effect of adaptation is reduced and the sensitivity of the animal increases. This effect of sensitization lasts for many minutes or even hours depending on the strength of the stimulus and represents a form of short-term memory.

In both cases, the change in the reaction of the neuron is a consequence of varying the intensity of the mediator release in its synapses.

It is believed that in mammals, a special role of modification and subsequent memorization of altered neural networks is played by the hippocampus – a special area of the cerebral cortex.

Other memory mechanisms are also discussed.

So, pulses wandering in the neural circuits, delayed on long transmission paths or returning again to the neuron via closed circuits of neurons can maintain a state of excitation or inhibition in the nerve center.

Other researchers suggest that the long-term preservation of the modified state in the nerve cell is based on the change in the structure of the proteins that make up the cell (possibly also the glial cell proteins).

Since the program space of external programs with objects in it, such as food or sexual partners, is subject to constant changes, it requires constant monitoring of these changes by sensory OMs, analysis of these changes and also correction of effector, motor programs in accordance with the new sensory information.

In other words, a new modified adequate motor complex is created for the new sensory situation. In general, this implies modification and updating of the corresponding local neural network.

In fact, this adaptation of the external program to the new conditions largely corresponds to the process of learning considering in the literature on the organism behavior.

The initial stage of training is reduced to the method of trial and error.

It should be noted that the participation of the hippocampus both in the formation of emotional reactions and in the processes of long-term memory suggests, for example, that the emotionally marked successful result of the modified program contributes to the memorization of the configuration of the corresponding updated neural network.

When learning, the memory mechanisms that are associated with the change in the efficiency of synaptic contacts depending on the frequency of signals passing through them considered in the previous section are used. Neural networks, which are rarely used, weaken and disappear, while frequently used ones become more efficient and available for use. In general, a well-known proverb works: "repetition is the mother of learning".

Improvement of External Programs at the Highest MO Representatives

The most important direction of multicellular evolution becomes the improvement of external programs, because the solution of the main task of species survival – the supply of resources, depends on their successful implementation.

As a result, all three external maker blocks: sensory, analytical and effector (muscle) ones have reached the highest level of development in the highest representatives of mammals.

Mammal sensors register large situational patterns with high spatial and intensity resolution for all modalities of external signals – optical (vision), sound (hearing), chemical (smell, taste).

The predator's analytical networks determine with high accuracy the position, speed and acceleration of the victim, and then select and use adequate combinations of muscle effector makers to capture the victim that requires a honed development of spatial-temporal situational representations and algorithms of their use.

Finally, the structure and functioning of muscle makers have reached a high degree of perfection, allowing to combine the accuracy, speed and coordination in the performance of their motor programs.

MAN

Brain Evolution and Hand Motility – Basis of Homo Programming Appearance

A new evolutionary human stage began on the way to further improvement of external programs. The rapid deterioration of the climatic and resource conditions of the environment in the Primate habitat area required from them a quick, and, consequently, quite local and at the same time-effective evolutionary response to this challenge.

And such an answer was found, first, in the form of a local change in the brain and, in particular, the creation of structures that provide opportunities for independent, volitional external programming.

These structures largely coincide with the "language" areas of Wernicke and Brock, because the language is the necessary symbolic base for the creation and modification of volitional programs (VPs).

In all languages, the basic words refer to such program components as the makers and objects (nouns), maker actions on objects (verbs), the site – place of the actions (noun, adverb), as well as features and characteristics of the components (adjectives, numerals, adverbs). For example, color, size, hardness of makers and objects, speed and intensity of action, geographical location and usability of sites, etc. In addition, significant sections of the "program dictionary" are evaluations of programs and their components, often represented in the form of antonymic pairs: good-bad, ours-not ours, expensive-cheap, fast-slow, etc.

Grammatical constructions composed from words and symbols, including sentences and messages, are programs, their scenarios and fragments.

In parallel with the appearance of the neuron-network analytical apparatus that supports the thought process work with VP, the effector hand muscular apparatus necessary for both the creation and use of the VP makers (tools) was developing.

Thus, for the first time in evolution, there was an organism with unique opportunities independently to make and carry out external programs. A man can deliver program task, ask the result, select and, if necessary, to make the makers and objects for it, to ensure energy and site for its implementation.

The emergence of the formed "Homo Programming" can be correlated with the era of Cro-Magnon man, when independent human activity was already fully manifested in all external programs, from hunting to housing construction.

Social Nature of Human Communities – ASVP Occurrence

It is significant that these VPs were mostly designed and implemented collectively, which led to the development of the tribal community.

Own inherent autonomous system of volitional programs (ASVPs) independently developed and at the same time carried out in the Cro-Magnon community, which was a socially geographically localized.

ASVP is the essence of an independent, self-sufficient human community capable of independent evolutionary development.

Manifestation of sociality in ASVP is essentially like what is observed in a MO.

Part of the resources a person takes away from their personal programs and spends them on participation in collective programs ASVP, receiving in return social benefits from the possible use of collective results. Naturally, the collective results of ASVP are fundamentally higher in terms of the level and range of those that could be achieved by the individual, performing only their personal programs.

Modern ASVP is identical to the state, with the exception of archaic tribal communities that have survived in hard-to-reach regions.

Personal and Collective Volitional Programs

Personal programs are made, modified and executed by each person individually and in turn are divided into internal individual and external – components in the collective programs.

Collective programs consist of external personal programs and other collective programs.

In personal individual, VPs can be identified basic, which include nutrition, reproduction, programs that provide homeostasis in terms of housing, energy, clothing (shoes) and sanitary, health (treatment) programs.

Basic collective programs ensure, first and foremost, the implementation of basic individual programs, providing individuals with food, energy, shelter, clothing, and executing sanitary, health and reproductive programs. A large group of collective programs provides all the collective programs, including these programs themselves by makers, energy, sites and objects, which fully corresponds to the group division of program subsystems in the cell.

In addition to the above effector production programs, which should also include transport programs, the most important role in the ASVP is played by management and IPs.

The structure of the management system of the ASVP includes management programs of different levels for both collective and personal programs.

Tactical goal is aimed at maintaining system stability in response to current deflecting influences from both external and internal factors.

And strategy is focused on increasing system resistance to these impacts through evolutionary development.

The main role of IPs in ASAW as in other PSs is to ensure reproducibility and modification of all programs and to refine the system's response to situational changes.

IPs, which include radio and television broadcasts, various media on paper: newspapers, magazines, books and of course the Internet information on a wide variety of programs: on the progress and conditions of their implementation; about modifications and new developments of the programs themselves and specifically of performers, objects and sites.

Social Allocation of ASVP Resources

Programs that implement two mechanisms – market and distribution, play an important role in the social allocation of goods and services produced by collective programs.

The market is a natural or virtual warehouse of finished products (goods and services) offered by the buyer to the consumer, on the conditions prevailing in the market at the time of sale and determined by the ratio of buyer demand and manufacturer offer. That is, when using the market mechanism, the interaction of production – consumption programs is regulated exclusively by the feedbacks between these programs in the absence of external control.

Money is an unambiguous and universal expression, the embodiment of the sale conditions, i.e., the cost of goods. Being the universal equivalent money allows you to estimate the cost of any goods and services.

As we know, each program requires resources: makers (in our case – labor force and tools), objects (semi-products, raw materials), sites and energy. In a market system, all resources are goods and services supplied to the market by other programs. In the case of labor, it is a market offering of personal individual programs. The equivalent of all these resources is money.

Thus, money, or rather a system of financial programs that organize and regulate money flows, is the basis for the functioning and development of the entire ASVP when using the market mechanism.

In case of implementation of the distribution mechanism, the consumption of goods and services is controlled by special volitional, command, administrative programs, bypassing the market mechanism.

Accordingly, in this case, the functioning and development of the ASVP as a whole are also implemented by administrative means.

The state with market management ASVP is designated by the term "market state", and its antipode with volitional, command management – "administrative".

The ability to manage a program, PSs, and whole an ASVP is usually called power, which can be personal, authoritarian or collective.

At the market control, the power is in the hands of a persons or groups with significant money resources and managing the financial programs.

The management of the administrative PS is the power in administrative state.

It should be noted that there are currently no states in the world with unambiguously market or administrative systems of ASVP management. Mixed management systems work usually with a predominance of one of them.

Management in ASVP

The following methods are used to manage programs and their systems.

First, the intensity of their implementation increases or decreases by changing the number of performers and co-executors. For example, the number of workers in the production and, or the number of machines changes. At the same time, working areas, volumes of purchased raw materials and components can change accordingly.

Outdated programs stop and up-to-date or modified programs run. Accordingly, there is a constant demand for developing and scientific programs that prepare modified and innovative programs.

Compliance with the rules of program execution, the so-called production discipline, as well as the interaction of the program components included in the subsystem is maintained and monitored.

To implement the management of ASVP, resources for the restructuring and updating of managed programs are needed. In most cases, such resources are both money and material component: executors (makers) for managed programs, their objects, new or modified sites of their execution.

The administrative part of the management is carried out as follows.

Formed and monitored plans to achieve programmatic results within a specified time, instructions and regulations for the scenarios implementation and their implementation conditions, etc.

The structure of the ASVP management system correlates with the geographical structure of the state, which has different levels: federal, regional (provincial), district and municipal. The higher the level of the structure, the larger the objectives and strategic long-term goals it sets and monitors.

In addition, there is a sectoral management structure by type of goods and services produced, which is also built on a hierarchical basis.

At the same time, the principle of large-scale division of the industry into ministries and companies, plants and enterprises, divisions and departments is primarily determined by the convenience and management efficiency of the relevant program subsystems.

Coordination and Synchronization of Programs

Harmonization between programs and their systems plays an important role in the management of ASVP and its structural components.

This alignment is necessary between different collective, collective and personal programs and individual personal programs.

Most often, the need to coordinate the implementation of separate, especially collective programs are caused by their mutual claims to the use of resources, both monetary and material.

Widespread controversy over claims for different kinds of resources between personal programs of relatives and neighbors take place.

Finally, one of the most important points of the optimal existence and development of ASVP is the minimization of contradictions between collective and personal programs. As already noted, an individual's PS consists of personal internal and external programs. The internal program first provides functioning of an individual organism and members of his family. The external program is a component of the collective program in which the individual works and serves as a source of resources for his internal program.

The main contradictions between internal and external programs that require resolution are related to the need to allocate time and energy costs between them, and thus between personal and collective programs.

Courts of different specializations and instances implement special conciliation programs in the most sharp and difficult situations.

The competence of the territorial (geographical) authorities includes a wide range of issues.

These are the issues of functioning and development of collective programs: consumer, procurement, organizational and information. The relevant departments of local and federal government are responsible for these matters.

Further, it is the control and regulation of the supply of personal programs with everything necessary from food to energy resources and from medical to leisure and entertainment services provided by the managing units.

They also include social programs: payment of pensions to the elderly and disabled, various benefits to the needy segments of the population, implemented by the departments of the pension fund and social protection of the population.

The regulations of disputes between different programs and the maintenance of public order are important management functions. This is done by an extensive PS of the Ministry of the Interior, the Prosecutor's Office, the courts and the penal system.

Finally, a tax system collects money in the form of taxes from individuals and businesses to finance the management programs.

All the above program directions of state ASVPs represent internal subsystem. However, there are also external ASVP programs managed mainly by federal agencies. These include, foremost, the relationship programs with external ASVPs.

Types of Management Regulatory Programs

In addition to the two management systems discussed above, administrative and market, mercantile, there is another moral system that includes ethical philosophical theories and religion.

A significant part of this system is devoted to the relationship of personal programs of different individuals and personal programs with collective programs. We are talking about the rights and obligations of the parties in this relationship, but not in the rigid paradigm of the established game rules of the administrative or mercantile systems but using the conscious application of the principles proposed by this system.

Briefly, these principles are summarized in two: prohibitions to commit acts that violate or terminate partnership or neighboring programs and incentives to act to promote the functioning and development of these programs.

The moral system refers to the practical human experience, and to his conscience.

As in other PSs, the most important part of the ASVP management control programs is the analysis of signals by which the control decision is made. The important criteria of the analysis are harmful/useful for a particular program and, in general, for the PS within which the analysis is performed.

Such criteria are used in the analysis of market and administrative (legal) management. Many administrative signals have the meaning of an unequivocal prohibition or encouragement and do not require analysis.

In the case of a moral system, less defined (unambiguous) criteria are used: good/bad, moral/immoral, noble/ignoble.

Evaluation noble, moral act is often used for the case when a moral decision is made even in spite of mercantile and sometimes administrative estimates.

It can be summarized that the ASVPs consist of subsystems of personal and collective programs, managed and regulated by three types of regulatory programs: administrative, mercantile and moral.

History of Evolution of Volitional Programs

One of the evolution purposes of all autonomous program systems (ASPs), including ASVP, is to increase stability to the external and internal influences rejecting system from a regular state.

The capacity of the PS to respond to deflecting imbalances is determined by the level of material and energy resources and the degree of perfection of the program management tools and their interaction and coherence and the evolution is to increase this level and the degree of management.

The increase in the resource intensity of programs is due both to the increase in the supply of resources and as a result of more economical use of them by the program.

The evolution of personal and collective programs is coordinated, albeit with its own characteristics.

The tasks of the evolution of personal internal individual programs are to increase the stability of homeostasis maintenance systems, by improving clothing and footwear, housing and household appliances and objects. In order to increase the body's resistance to diseases and higher physical and mental stress become more diverse and effective food, tools and methods of their preparation and storage. For the same purpose, medical and health programs are being improved.

The development of consumer personal programs inevitably leads to the evolution of collective programs that provide these programs. Accordingly, the energy consumption of the entire PS in all its forms is growing, which requires intensification and improvement of energy programs.

Organizational and IPs are becoming increasingly important because of the evolutionary development, expansion and complication of personal and collective programs. The means of transport and communication improve more intense. The role of IPs required to identify and resolve contradictions between more and more numerous and diverse programs is rapidly increasing

Thus, the evolution stimulated by personal programs covers the entire ASVP.

The benefits derived from evolution are spread and consolidated through the exchange of information, using both administrative and market mechanisms, both within and outside the ASVP.

Thus, with the development of information, communication and transport programs and the globalization of market and administrative management programs, the pace of the ASVP evolution is increasing dramatically.

In general, the rate of ASVP evolution, since its beginning several tens of thousands of years ago, is exceeding the rate of classical gene evolution by many orders of magnitude.

Modern man, almost unchanged since the Cro-Magnon era (on the genome), every few decades is in a new structure ASVP.

The Role of Science

As already noted, the history of mankind is the evolution of VPs, or rather ASVPs, and the main direction of this evolution is the improvement and generation of new VPs. Until the end of the 16th century, significant evolutionary events, dramatically accelerating it, occurred more or less randomly when resources accumulated in the most "advanced" ASVPs.

And only from the 17th century evolution began actively developing a special mechanism of its acceleration – science.

Science is a specialized IP subsystem of ASVP, which aims to produce, create and organize new information for the development of all ASVP programs, including the scientific programs themselves.

At the same time, the ASVP management programs constantly increase the share of resources allocated to this subsystem that corresponds to the general trends of accelerating the evolution of individual ASVP in the situation of their increasing competition.

To date, the scientific program subsystem consists of two parts – applied and fundamental.

The applied one defines the modernization of all consumer programs, including food, household, construction, medical and entertainment programs, etc., and also information and transport programs that are especially important at the present stage.

Fundamental science prepares the scope of work for applied science.

The Scientific Method from the Standpoint of the Program Approach

Some elements of the methodology of fundamental science can trace to the example of physics.

The basis of physics is an experiment.

An experiment is the registration of a reproducible reaction of an inanimate object to a regulated program of experimental influence.

Indeed, a competent classical physical experiment involves the mandatory fulfillment of the following conditions:

- Selection of an object that allows observing its parameters.
- The choice of a certain modality and regulated in space and time impact on the object.
- Elimination or at least suppression of all impacts and phenomena caused by them that not relate to the studied.

Cleaning the experimental program allows one to isolate the predictable, in fact, the program manifestation of the lifeless object that defines as a regularity or a law. Description of the real object behavior occurs when adding to the ideal "program" laws of additional third-party impacts on the object is often stochastic in nature, for example, friction in mechanics.

We can say that physics forms the program perception of stochastic processes of the inanimate world, using a regulated program experiment.

In general, the scientific understanding of the inanimate (non-program, stochastic) world is a set of reproducible reactions of its objects to experimental well-regulated exposure programs.

At the same time, the rules of the program of physical experiment, based on the concepts of time, space, causality, are traditionally attributed to the stochastic object of study, and then to the whole stochastic inanimate world.

About the World Picture

The modern picture of the world is physics-centric, which is based on the laws of the inanimate world: ideas about the elements of matter formed by particles and fields, and their interaction.

The basic methodological approaches to the study and knowledge of the world order: the concept of time, space and cause-effect relationships a priori assumed inherent properties of the inanimate world.

The living more complex organized part of the world is supposed to originate from the inanimate part of it, that is, retains all the regularities of the structure and behavior of this inanimate world.

If we accept the proposed in this book program-information concept of the living world, then, creating an overall picture of the world, we have to take into account the following circumstances.

The cardinal difference between the ordered, predictable, reproducible and, at the same time, evolutionarily developing living world from the stochastic inanimate one leads to the principal irreducibility of the first, as to a certain, higher phase of the development of the second.

Certainly living world bases on the main laws of the non-living world – particles, fields and interactions between them. However, the manifestation of these laws in the living part of the world is radically different in terms of the implementation of interactions between the objects of this living world.

Instead of stochastic arbitrary interactions of equivalent objects of the non-living world, leading to random processes, in the living world interactions are ordered and regulated in time and space and are performed between unequal objects, one of which is an active maker performs actions in relation to another passive object.

The new quality of behavior of living objects cause by the presence of an additional IS in the living nature, which determines both the space-time regulation of interactions and their causal relationships.

It is with the help of this system that a person, following all living organisms, learns the world, including its inanimate part, and interacts with it.

Thus, in the proposed picture of the world, the main methodological, temporal, spatial and causal approaches to the study and cognition of the entire world represent as derivatives of the living part structure, namely, its inherent IS. These approaches are not a consequence and manifestation of the laws of the inanimate world. Moreover, since all the basic laws of the inanimate world have been studied and are being studied using the "subjective" IS of the living, they bear the features of this "subjectivity". This in no way calls into question the validity of these laws, although it allows for clarification of their wording and interpretation in some cases.

Thus, the program-information structure of the living world is brought to the basis of the world picture along with the general physical material structure – particles, fields and their interactions.

Mechanisms of Volitional Programming

Opposition of communicative and mental functions of the language is no longer relevant if the program approach is used. In both cases, the subject is external programming: compiling programs from components – makers, their actions and objects, the choice of sites and conditions for their implementation, as well as memorizing and subsequent use of ready-made, proven programs and, if necessary, their modification.

Language is the basis of personal volitional programming, collective programs and of course all communication processes that permeate the functioning and development of an ASVPs.

The basic element of VPs is the information pair (IP): the signal mark – word and the corresponding image.

Usually, the pattern of the word causes an image pattern, which, as we can assume, is performed by a certain local network of neural deciders, the inputs of which are fed by the nerve impulses of the pattern of the word, and the output generates pulse pattern of the image.

The memory "library" section of VPs in the human brain can be divided into two parts: the main and auxiliary.

The main library contains specific program information. These are the programs themselves, their structures and scripts, as well as their elements: makers, objects, actions, results, sites; characteristics: time and speed of execution, efficiency; conditions of execution: prohibitions of execution or its stopping, favorable situation for execution, etc.

The auxiliary library collects and structures preliminary, overview, general information about the human environment and the world. The person needs this information for the correction and modification of existing programs, creation of new and, very importantly, to reconcile and correct cooperation their personal programs with the neighboring surrounding ones.

Two control programs: manipulative and stimulating manage the mechanism of volitional programming and, in particular, working with libraries.

Manipulative programs work with all, without exception, programs (including themselves), namely, extract them and their elements from the library, provide their virtual execution, carry out their modification, evaluate the results of modification and monitor their real execution, and so on.

Stimulating ones, occupying the highest level in the hierarchy of control managing VPs, decide which of the programs to be run and processed by the manipulating program, for what, in particular, they use a system of program schedules, diaries, calendars that are an important part of information libraries.

Here we can introduce *the concept of thinking as a set of information operations and transformations with images and symbols of all program elements and scenarios of VPs, as well as of their features and assessments necessary for the reproduction, modification and compilation of new programs, what makes up all the conscious activity of man.*

Accordingly, *human consciousness (also self-consciousness) – self-controlled work with VPs of their choice, reading, execution, modification or compilation of new ones.*

Features of Volitional Information Systems

The main feature of IPs in humans is the appearance of a secondary internal signaling system, essentially a language system, the basis of which is made up of pairs: a word (symbol) – an image of a program element, the program itself and their signs, and ratings.

This information is stored in the memory in such a way that it forms associative hierarchical networks, allowing one to produce the image grouping into different blocks by their content, for example, to choose red color fruits, the fastest cars or makers for processing wood products.

All this reminds us of modern Internet search engines. Do not be surprised, because they are made in the human likeness and are designed for the brain network of VPs.

In addition to the library of the secondary signal system, the essential novelty of the human IS are new programs that implement work with this library, primarily reading and writing information as well as manipulation of symbol – image pairs for program evaluation and modification.

General management of the work with VPs and stimulation of this work are also essential.

Apparently, these programs are implemented in the "human" brain departments of Wernicke and Brock, as well as in the "will zone", detected in MRI studies. Possibly, the symbol-image library is also located there.

It can be assumed that the basis of the structure and functioning of these additional parts of the human brain are neural networks similar to those which provide transformation of information patterns in MOs.

So far, we have talked about the brain part organization of individual, personal VPs.

However, in the very early period of the existence of Homo Programming formation, creation collective IPs that helped to work with the program library, modify and create new programs also began.

This could be, for example, the development of criteria for assessing the quality of makers – tools or methods of drawing up a program of making a maker.

The basis of this first collective IS was oral language.

The emergence of the written language was a huge leap in this area, as there was a reliable base for the creation of a collective program library, and for editing and creating new volitional programs.

In our time, the role of collective IPs has increased immeasurably, and individual average social person cannot take a step without information tools for collective use or individual information gadgets made by numerous corporate teams.

Directions of development of information technologies are the same as tens of thousands of years ago at the beginning of the evolution of VPs. These are storage devices, means for transformation of information patterns and arrays, means of information exchange between subjects.

EVOLUTION AND PROGRAM APPROACHES

From the point of view of the program approach, evolution is an adaptive change of programs and the whole program-IS of the organism, i.e., its protype on a long (more than tens of thousands generations) time interval.

The essence of the protype evolution was the transformation, change in cellular organization, consisting in multicellular in specialized differentiation of all cells and the creation of a mechanism of social exchange of a specific cell product for centrally supplied food and oxygen.

In humans, this transformation affected only a few areas of the brain, which led to the global changes of the protype.

It should also be noted that, starting with the advent of multicellular, it was external programs that underwent more and more evolutionary changes, since the successful fulfillment of the vital tasks of survival – nutrition and reproduction – depends on them.

The evolutionary spurt of man was entirely due to the emergence of fundamentally new opportunities for independent volitional creation of external programs.

Moreover, the IS that organizes the body's PS and at the same time is its organic part (which is why we often say "program-information system"), embedded in the cell and developed in MOs, "went outside" in the conscious VPs of a person.

For understandable "hereditary" reasons, the tasks, organization, composition of program elements in humans have not fundamentally changed.

Information support for regulated reproducibility and adequate adaptability of a PS is provided in a cell automatically in a "hardware" way, i.e., the specific composition of the corresponding makers created within this system.

While a person has to himself, meaningfully, create new program elements, compose and run a program, evaluate its result and, if necessary, modify it, thereby realizing his many times accelerated evolutionary adaptability.

Naturally, for this, a person had to master and develop all the informational principles inherent in the cellular program of organizing all living things. And when we say that the roots of mathematics can, and perhaps should be sought in the cell, this should be less surprising than an explanation of their appearance by insight of unknown origin.

Provisions of the Program Approach

* Program organization is common to all evolution stages – unicellular, multicellular and human.

* All actions that take place inside a living organism or perform by the organism itself in the external environment are the result of a predetermined, orderly transformation of the properties of a passive object by an active participant in the interaction, the so-called maker or its agent.

* A set of such regulated actions, leading to the result: the division of the object, the assembly of the object from the component parts, the movement of the object or its change is called the program.

* Apart from the maker and the object, the components of the program are: the site – the place of the program execution, and the energy necessary for the maker to perform an action on the object.

* Makers inside the cell are proteins, ribonucleic acids and the complexes created from them.

* The behavior of the maker includes trial contact with the object, object recognition, the formation of a working maker-object contact, the implementation of the program action of the maker over the object, the separation of the maker and the object modified as a result of the object action.

* Maker is specialized for performing certain action on a specific object.

* All elements of the program have a basic nature, common with the non-living part of the world, namely the system of particles, fields and their interactions.

* The stochastic process of the non-living opposed to the ordered program is a set of equivalent interactions of two arbitrary bodies and objects occurring in an indefinite time in an undefined place.

* The fundamental difference of life is the presence of an information system that transforms the basic stochastic interactions into ordered, regulated program actions of the maker in relation to the object.

* The central element of the information system is the pair: information mark – information domain of the maker.

* Sources and/or localization of marks can be inside or outside of an organism.

* The result of the information interaction of the mark with the maker domain is either a change in the activity of the maker, or generation of another information mark.

* There are three types of marks: instructive, identification and signal.

* Instructional marks contain information by which the components of all the makers are collected: proteins and RNA.

* Identification marks are used for identification of the object by the maker mainly.

* Marks-signals form signal program chains, often combined in information networks, for solving the problem of synchronized program management at an adequate response to the changes in the internal and external situation of the cell.

* Interaction of the signal mark with the maker domain forms information chains starting with the primary signal mark and ending with the signal mark that controls the activity of the effector program maker. These chains form information networks.

* Cellular programs form hierarchical structures consisting of actions – steps, and also component, composite and integrated task programs that are carried out by properly structured makers.

* The system of task programs contributes to the main goal achievement – to ensure their reproduction and functioning in the changing environment.

* Task programs are divided into two groups: effector and information.

* The main effector programs are those that provide the creation of makers, extraction and storage of energy, extraction and preparation of objects, creation of sites, maintenance of homeostasis inside the organism, counteraction to the accumulation of stochastic defects, reproduction.

* Signal and control information programs provide a coordinated, synchronized work of the organism program system, necessary for the timely passage of the organism phases and adequate response to changes in internal and external situation.

* Essential elements of information system are information makers – **deciders** which take place in the information chains between sources of signals, receptors and chain endpoints launching the targeted effector maker.

* The p53 protein is an example of a comprehensive decider defining adequate passage of the cell cycle and its response to external and internal impacts.

* The main control mechanisms of programs effect on their playback frequency.

* The playback frequency is determined by the concentrations of the makers, their objects and energy particles on the program site primarily.

* The maker concentration control is crucial.

* The required maker concentration is provided by two main programs – maker producing and ready-made maker transporting on their site.

* The most important of them is the first, producing program, and in it, the transcription program.

* There are many mechanisms for regulating the frequency of this program in the cell using combinations of different regulatory proteins interacting with the promoter part of the transcribed gene.

* Marks form composite information patterns, which in turn can form more complex patterns of higher orders of hierarchy, as observed, for example, in gene instructions and situational signal patterns.

* Possible operations with information marks:

- The convolution of the pattern into a simple mark or a simpler pattern
- The disclosure of the higher hierarchy mark into a pattern
- Mark and pattern recognition with subsequent reaction in signal generation
- Mark and pattern memorizing

* The information network formed by deciders can perform with signal marks the logical operations: negation, conjunction, disjunction, and to solve with them complex logical problems for making reactionary decisions that are appropriate to the current situation.

* Information system (IS), which includes information programs and information actions in effector programs, is used for reproduction, interaction and synchronization of programs, as well as for realization of cell cycle phases and adequate reactions to changes in internal and external situations.

* The space-time and causality concepts permeate the entire program cellular organization and they are included in the IS.

* To achieve evolutionary goals, the necessary and sufficient system of single-celled programs should include the following task groups: maker, resource, transport, compartment, homeostasis, signal, control, reproduction, anti-stochastic, protective and evolutionary ones.

* Intermediate forms between non-living matter and the first ancestral cell, which already has a complete program set, are not capable of sustainable multiple reproduction necessary for the evolutionary process.

* Multicellular organism (MO) is a social association of specialized cells that perform their specific programs to provide general task programs of MO.

* Basic social programs of MO maintain life activity of all MO cells, by supplying them with necessary components of nutrition and oxygen, as well as implementation of immune protection from pathogenic microorganisms.

* Cells, spending their resources to fulfill their specific social programs, have the benefit of the guaranteed resource supply and immune defense that is evolutionary perspective, compared to unicellular.

* Specialized social makers perform social cell programs.

* Cells containing social makers are called differentiated, and implementation of the social maker formation programs is called differentiation.

* A cell within the MO has its own specific makers, which essentially execute external programs in relation to the cell itself, but internal within the whole organism.

* The rest nonsocial makers perform internal cell programs, often referred in the literature as the term "household".

* Social makers form a multi-level system to perform common MO task programs.

* It includes the molecular "elementary" makers (EM), makers – cell or cell makers (CM), makers of tissue or tissue makers (TM), makers of organs or organ makers (OM), makers of the system or system makers (SM).

* The most important social programs for the exchange of material resources and information signals between the makers of all levels are carried out by blood and nervous systems of MO.

* Cells receive both "material supply" in the form of nutrient molecules and oxygen, and signal marks for managing their programs from the intercellular space filled with interstitial fluid.

* Cells emit the products of their social programs and signal marks for the nervous and humoral systems in the same interstitial fluid also.

* Being a social formation of specialized cells, MO inherits cellular information principles.

* The MO reproduces a similar structure of network formation from signal information chains, which include three types of links. There are receptor links generating primary signals, analytical decision-making links and effector links sending reactive signals to target effector makers.
* The main difference between the cell and MO networks consists in the arrangement of links of such chains. If in a cell chain links are the elementary protein makers, then in MO they are neurons with a fast electrical propagation mechanism of pulse membrane potential.
* Neural information chains form information networks. Intermediate, analytical neurons, n-deciders (by analogy with protein deciders) play a critical role in the formation and functioning of the neural networks.
* The n-deciders can carry out all the basic logical operations: conjunction, disjunction, inversion, implication and to estimate the result effects may be involved analog arithmetic operations of summation and subtraction.
* The greatest development of network information system of the MO, especially of the highest mammal representatives has gained in connection with the growing diversity and evolutionary significance of external programs.
* As a result, all three external program blocks: sensory, analytical and effector have reached a high level of development.
* The main change in the human organism, which occurred during the evolutionary period from Australopithecus to CRO-Magnon, is the appearance of the brain fields of Wernicke and Broca, responsible for the language perception and processing.
* Both these structures are expressed quite clearly already on the endocranial cast of fossil skulls of Homo representatives.
* The appearance of these fields was the basis for the emergence of a new type of external programs – independent volitional programs (VPs).
* In all languages, without exception, the basic words denote such components of programs as makers and objects (nouns), the actions of makers on objects (verbs), the site – the place of such action (nouns, adverbs), as well as the features and characteristics of these components and their assessment (adjectives, numerals).
* Grammatical constructions composed from words including sentences and messages are program scenarios and the actual program or its fragments.
* Speaking to himself or aloud the relevant words and messages composed of them, a person at any time of his own will, desire can virtually or really play any pre-memorized program, as well as modify the old program and create a new one from the known memorized components.
* The basic element of VPs is the information pair – word (symbol) and the corresponding image.
* It is possible that the word pattern triggers, generates the pattern of the image, what, as can be assumed, is performed by a certain local network neural decider, the inputs of which are supplied with nerve pulses of the word pattern, and the pulses of the image pattern are generated at the output.
* The memory section of VPs in the human brain can be conditionally divided into two parts: the main and auxiliary libraries.
* The main one contains specific program information: the programs themselves, their structures, scenarios, elements, and conditions of implementation.

* Preliminary, overview, general information about the human environment and the world is collected and structured in the auxiliary library.

* From the programs that control the volitional programing mechanism and, in particular, the work with libraries can be highlighted conditionally two: manipulative and stimulating.

* Manipulation programs work with all, without exception, VPs (including themselves), namely, extract them and their elements from the library, provide their virtual execution, carry out their modification, evaluate the results of modification and monitor their real execution.

* Stimulation programs, occupying the highest level in the hierarchy of VP management, make decisions about which program is to be launched and processed by the manipulative program. For selection of launching programs a system of schedules and calendars in information library is used.

* Thinking – a set of information operations and transformations with images and symbols of all program elements and scenarios of VPs, as well as their features and assessments necessary for the reproduction, modification and preparation of new VPs.

* Human consciousness (self-consciousness) – self-controlled work with VPs: their choice, reading, execution, modification and compilation of new ones.

* During the same evolutionary period of the establishment of VP brain add-ins, effector arm muscular system to create and use makers (traditionally called tools) for the programs is developed.

* Thus, for the first time in evolution, an organism with unique capabilities to independently compose and execute external programs arose.

* The emergence of the formed "Homo Programing" can be correlated with the era of the CRO-Magnon man, when the independent human activity was already fully manifested in all external programs, from hunting to housing construction.

* These programs were mostly designed and implemented collectively, which led to the development of the social community with the tribal structure.

* We can talk about the formation in this period of Autonomous Systems of Volitional Programs (ASVPs) – social independent, self-sufficient human communities, capable of independent evolutionary development.

* Manifestation of sociality in ASVPs is essentially similar to what is observed in an MO.

* Part of the resources a person takes away from their personal programs and spends on participation in collective programs ASVP, receiving in return social benefits from the possible use of a variety of collective results.

* Personal programs are made, modified and executed by each person individually and in turn are divided into internal individual and external components in the collective programs.

* Collective programs consist of external personal programs and other collective programs.

* Basic personal programs include nutrition, reproduction, programs that provide homeostasis in the part of housing, energy, clothing (shoes) and sanitary, health (treatment) programs.

* Basic collective programs ensure the implementation of basic individual programs and all collective programs, including these programs themselves, by makers, energy, sites and objects.

* Transport, management and information programs are among the most important programs that provide the social structure of the ASVP.

* The organization of social distribution of goods and services produced by collective programs is carried out by programs that implement two mechanisms – market and distribution.

* In the management of the ASVP, three types of regulatory programs are used: administrative, market (mercantile) and moral (religious).

* The main role of information programs in the ASVP, as in other program systems is to ensure the reproducibility and modification of all programs and to implement the system's response to situational changes.

* One of the goals of the ASVP evolution is to increase the resistance to external and internal influences, which deviates the system from the standard state.

* Personal program driven evolution embraces the entire autonomous system of VPs.

* The rate of ASVP evolution since its beginning several tens of thousands of years ago has exceeded the rate of classical gene evolution by orders of magnitude.

* Approximately from the 17th century, evolution began to actively develop a special mechanism of its acceleration – science.

* Science is a specialized information program subsystem of ASVP, which aims to produce, create and organize, systematize new information for the development of all ASVPs, including the scientific programs themselves.

* To date, the scientific program subsystem consists of two parts – applied and fundamental.

* Application defines the modernization of all consumer programs.

* Fundamental part prepares front for applied science.

* The basis of fundamental science an experiment, for example, a physical one is carried out according to a research program that allows to isolate the reproducible behavior of an inanimate object under the influence of a strictly regulated effect.

* Such essential program (ideal) behavior of the original stochastic non-living object is defined as regularity or, more generally, a law.

* A world picture that takes into account the evolving programmatic approach should include two important circumstances in its concept:

- Irreducibility of the living world to a certain, higher phase of the development of the inanimate
- Program (from the living world) origin of the principles used in the knowledge of the inanimate world, primarily space-time and causal principles.

* The main methodological, temporal, spatial and causal approaches to the study and knowledge of the world represent by derivatives of its living part structure, namely the information system inherent in it. These approaches are not a consequence and manifestation of the non-living world laws.

GENERAL CONCLUSION

Life – evolution of program-information organism systems.

Common features of the three evolutionary stages: unicellular, multicellular and human

1. ***Program-maker structure***
2. ***Information system***
3. ***Goal structure: reproduction, adaptation, evolution***

Reading List

Bruce Alberts, Alexander Johnson, Julian Lewis, Martin Raff, Keith Roberts, Peter Walter and 4 more. 2007. *Molecular Biology of the Cell*, Fifth Edition. Garland Science, New York.

Jim Al-Khalili, Johnjoe McFaddens. 2015. *Life on the Edge: The Coming of Age of Quantum Biology*. Black Swan, London

Michael J.F. Barresi, Scott F. Gilbert. 2021. *Developmental Biology*, Twelfth Edition. Sinauer Associates, New York, Oxford.

L.A. Blumenfeld. 1981. *Problems of Biological Physics*. Berlin, Springer-Verlag.

L. Brillouin. 1964. *Scientific Uncertainty and Information*. Academic Press, New York.

D.S. Chernavskii. 2000. The origin of life and thinking from the viewpoint of modern physics. *Physics-Uspekhi*. 43, 151–176.

P.M. Chumakov. 2007. Versatile functions of p53 protein in multicellular organisms. *Biochemistry* (*Moscow*). 72, N 13, 1399–1421.

Jerry A. Coyne. 2010. *Why Evolution Is True*. Oxford University Press, Oxford.

Peter J. Delves, Seamus J. Martin, Dennis R. Burton, Ivan M. Roitt. 2017. *Roitt's Essential Immunology*, Thirteenth Edition. Wiley Blackwell, New Jersey.

Donald A. Dewsbury. 1978. *Comparative Animal Behavior*. McGraw-Hill, New York

Lee Alan Dugatkin. 2009. *Principles of Animal Behavior*, Second Edition. W.W. Norton & Company, New York.

Gary Ferraro, Susan Andreatta. 2017. *Cultural Anthropology: An Applied Perspective*, Eleventh Edition. Cengage Learning, Boston, MA.

Thomas H. Frazzetta. 1975. *Complex Adaptations in Evolving Populations*. Sinauer Associates. Sunderland, MA.

Mark E. Furman, Fred P. Gallo. 2000. *The Neurophysics of Human Behavior: Explorations at the Interface of Brain, Mind, Behavior, and Information*. CRC Press, Boca Raton, FL.

David H. Hubel. 1995. *Eye, Brain, and Vision (Scientific American Library)*, Second Edition. W.H. Freeman, New York.

G.R. Ivanitskii. 2010. 21st century: What is life from the perspective of physics? *Physics-Uspekhi*. 53, 327–356.

Bruce M. Koeppen, Bruce A. Stanton. 2018. *Berne & Levy Physiology*, Seventh Edition. Elsevier, Philadelphia, PA.

Nick Lane. 2015. *The Vital Question: Energy, Evolution, and the Origins of Complex Life*. W.W. Norton & Company, New York.

Robert Lanza, Bob Berman. 2010. *Biocentrism: How Life and Consciousness Are the Keys to Understanding the True Nature of the Universe*. BenBella Books, Dallas, TX.

Clark Spencer Larsen. 2019. *Essentials of Biological Anthropology*, Fourth Edition. W.W. Norton & Company, New York, London.

Jon Lieff. 2020. *The Secret Language of Cells: What Biological Conversations Tell Us about the Brain-Body Connection, the Future of Medicine, and Life Itself*. BenBella Books, Dallas, TX.

David L. Nelson, Michael M. Cox. 2008. *Principles of Biochemistry*, Fifth Edition. W.H. Freeman & Company, New York.

John C. Nicholls, A. Robert Martin, Bruce C. Wallace, Paul A. Fuchs. 2001. *From Neuron to Brain*, Fourth Edition. Sinauer Associates, Sunderland, MA.

Addy Pross. 2012. *What Is Life? How Chemistry becomes Biology*. Oxford University Press, Oxford.

H. Quastler. 1964. *The Emergence of Biological Organization*. Yale University Press, New Haven, CT.

E. Schrödinger. 1945. *What Is Life? The Physical Aspect of the Living Cell*. University Press, Cambridge, England.

Gordon M. Shepherd. 1987. *Neurobiology*, Second Revised Edition. Oxford University Press, Oxford.

Charles Thaxton, Walter Bradley, Roger Olsen, James Tour, Stephen Meyer, Jonathan Wells, Guillermo González, Brian Miller, David Klinghoffer. 2020. *The Mystery of Life's Origin*. Discovery Institute Press, Seattle, WA.

C.H. Waddington. 2008. *The Origin of Life*. Routledge, London and New York.

Joanne M. Willey, Linda M. Sherwood, Christopher J. Woolverton. 2017. *Prescott's Microbiology*, Tenth Edition. McGraw-Hill, New York.

Index

A

Acetyl-CoA, 13
Adaptation, 109
Adaptive immunity (AI), 66
Adenosine triphosphate (ATP), 7, 14, 58, 101
  energy particles of, 5
  synthase, 4
Administrative management, 130
AEPs, *see* Anti-entropic programs
Aggregate organisms, 52–53
AI, *see* Adaptive immunity
AL, *see* Auxiliary library
Aminoacyl-tRNA synthetase, 10
Analytical blocks, 98
Anti-entropic programs (AEPs), 36, 160
Antistochastic programs, 33
Applied science, 133
ASPs, *see* Autonomous program systems
Associative learning, 109
ASVPs, *see* Autonomous system of volitional programs
ATP, *see* Adenosine triphosphate
Autonomous program systems (ASPs), 127, 172
Autonomous system of volitional programs (ASVPs), 113, 114, 131–134, 140, 175
  communication, 126, 127
  coordination and synchronization of, 171–172
  management of, 116–122, 170–171
  management regulatory programs, types of, 172
  occurrence, 168
  resources, social allocation of, 169–170
  synchronization of, 116–122
Auxiliary library (AL), 124–126, 133, 137, 148

B

Basic consumer programs, 128–129
Basic production of cell programs, pool of, 5–19
  makers, 5–11
  provision and energy programs, 11–14
  site reproduction, 14–15
  transport programs, 15–19
Basic social principles of cellular integration into single structure, 53
Bit, 47
Blood supply system (BSS), 54, 55, 71, 75–78
Brain evolution, 167–168
Breathing programs, 61–65
BSS, *see* Blood supply system

C

cAMP, *see* Cyclic adenosine monophosphate
Capillaries, 74
Carrier proteins, 19
Causality, 49, 143
Cell
  chemical laboratories in, 1–2
  definition of, 155
Cell cycle management, 29–32
Cell information system, 47
Cell makers (CMs), 162, 164, 165
Cellular evolution, 154–155, 161
Cellular makers (CMs), 54, 63–65, 70, 71, 75, 79, 86
Cellular program system (CPS), 3, 8, 19–29, 42, 45
  antistochastic programs, 33
  control programs, 20–24
  evolutionary programs, 33
  homeostasis, 33
  mitosis, 27–29
  protective programs, 33
  signal programs, 24–27
Central nutrition program (CNP), 59–61
CFM, *see* Complex fusion of makers
Channel-forming proteins, 19
CheA proteins, 34
CheB proteins, 35
Chemotaxis program, 46, 68
CheR proteins, 35
CheW proteins, 34
CheY proteins, 34, 35
CheZ proteins, 34
Circulation, 75
CMs, *see* Cell makers; Cellular makers
CN, *see* Command neuron maker
CNP, *see* Central nutrition program
Collective programs (CP)
  types of, 114–116
Colony-stimulating factors (CSFs), 67
Command neuron (CN) maker, 87, 88
Communication, 126–127
Compartment programs, 36, 160
Complete system of unicellular programs, 159–161
Complex fusion of makers (CFM), 106

Control programs, 20–24, 36
Covalent modification, 21
CP, *see* Collective programs
CPS, *see* Cellular program system
Cumulative or additive inhibition, 22
Cyclic adenosine monophosphate (cAMP), 24, 25, 52, 82

D

DCs, *see* Dendritic cells
Dendritic cells (DCs), 66, 69, 71
Deoxyribonucleic acid (DNA), 7, 8
Department of Foreign Trade, 119
DNA, *see* Deoxyribonucleic acid

E

Effector blocks, 98
Effector neurons, 86–87
  signal processing programs run by, 84–86
Elementary makers (EMs), 54, 61, 65–67, 69, 70, 77, 82–84, 162
EMs, *see* Elementary makers
Endoplasmic reticulum (ER), 14–18
Endothelial cells, 74
Energy programs (EPs), 2, 11–14, 28, 110, 129
Enzymes, 3
EPs, *see* Energy programs; External programs
Escherichia coli (*E. coli*)
  positive regulation, 8
  synthesis of β-galactosidase enzyme in, 23
Eukaryotic cells, 7
Evolutionary programs, 33, 37, 161
External programs (EPs), 51–53, 55, 56, 95–104
  formation of, 103–104
  at highest MO representatives, improvement of, 167
  inheritance of, 103–104
  program space, 97–98
  sensor, analytical and effector blocks, 98
  structure of, 98–99
  work scheme and interaction of blocks, 99–103

F

Flagella rotation program (FRP), 34
Food fragments (FFs), 55, 57, 59–61, 96, 97
FRP, *see* Flagella rotation program
Fundamental science, 133

G

Genotype, 154
Glycocalyx, 59
GM-CSF, *see* Granulocyte-macrophage CSF
GPCR, *see* G-protein coupled receptors
G-protein coupled receptors (GPCR), 25, 26
G-proteins, 25, 26
Granulocyte-macrophage CSF (GM-CSF), 67
GTP, *see* Guanosine-5′-triphosphate (GTP)
Guanosine-5′-triphosphate (GTP), 10

H

Hand motility, 167–168
HCl, *see* Hydrochloric acid
HCPs, *see* Humoral control programs
Heart, 72
Homeostasis, 33
  programs, 36, 160
Homo Programming, 111–113, 123, 144, 151, 167–168, 177
Human memory (HM), 123, 124
Human social paradigm, 114
Human volitional networks, emergence of, 111–113
Humoral control programs (HCPs), 79
Humoral system, 79–80
Hydrochloric acid (HCl), 59

I

ID, *see* Information domain
Identifier marks, 45
Image, 145–147
IMA, *see* Integrated maker association
Immune programs
  characteristics, 70–71
  implementation examples, 66–70
  tasks and structure, 65–66
IMs, *see* Intelligent makers
Informational organization of living, 2–3
Information domain (ID), 3, 48
Information network, 157
Information operations, 48
Information pair (IP), 145, 146
Information programs, 129–130
  task and structure of, 158–159
Information structure, 153–154
Information system (IS), 3, 43, 112, 123, 142, 151, 152, 159, 175, 177
  definition of, 49
  features of, 165
Information transfer, 126–127
Input sensor links, 83–84
Instruction marks, 45, 154
Integrated maker association (IMA), 162
Intelligent makers (IMs), 46
Internal programs, 55
  breathing programs, 61–65
  central nutrition program, 59–61
  immune programs, 65–71

nutrition program, 56–59
transport programs, 71–78
Interneurons, signal processing programs run by, 84–86
IP, *see* Information pair
IS, *see* Information system

K

Krebs catabolic cycle, 4, 13, 14

L

Lac-operon enzymes, 23–24
Language, and volitional programs, 144–145
LDL, *see* Low-density lipoprotein
Learning, 140, 166–167
Leisure programs, 115
L-isoleucine, 22
Low-density lipoprotein (LDL), 18
L-threonine, 22
Lymphocytes, 66

M

Macrophage cells (MCs), 66
Macrophages, 66
Main and auxiliary task programs, 55
Makers, 3–4, 41
intelligent, 46
production of, 5–11
programs, 35, 160
Management, control programs, 160
Management programs, 78–94, 157
humoral system, 79–80
memory of neural network, 93–94
nervous system, 80–93
Management regulatory programs, 172
Manipulative program (MP), 148
Market regulatory programs, development of, 131
MCPs, *see* Methyl-accepting chemotaxis proteins
MCs, *see* Macrophage cells
Memorizing, 166–167
Memory, 123–126
devices, 47–48
of neural network, 93–94
Messenger RNA (mRNA), 6, 10
Messengers, 46
Methyl-accepting chemotaxis proteins (MCPs), 34, 35, 68
Microcirculatory vascular bed (MVB), 72–74
Military programs, 131
Ministry of Economy, 120
Ministry of Finance, 120
Ministry of Health and Social Welfare, 121
Ministry of Internal Affairs, 119
Ministry of the Interior, 120
Mitochondria, 15
Mitosis, 27–29
Molecular chaos, xi
Moral regulatory programs, 131
MOs, *see* Multicellular organisms
MP, *see* Manipulative program
mRNA, *see* Messenger RNA
Multicellular organisms (MOs)
aggregate organisms, 52–53
basic social principles of cellular integration into single structure, 53
breathing programs, 61–65
external programs, 95–104
external programs at highest MO representatives, improvement of, 167
humoral system, 79–80
immune programs, 65–71
information system, features of, 165
internal programs, 55–78
management programs, 78–94
memorizing and learning, 166–167
nervous system, 80–93, 164–165
nutrition program, 56–61
program and information structures of, 104–110
program scheme of, 53–56
programs, management of, 163
programs, types of, 162–163
socialization, 51
structure of, 161–162
transport programs, 71–78
in volitional programs, inheritance and development of, 149–152
Multivalent inhibition, 22
Myeloid cells, 66

N

Nervous system (NS), 164–165
effector neurons, 86–87
input sensor links, 83–84
network structure, formation of, 87–93
neural chain links, 83
neuron activity, restoring, 82–83
neurons form NS signal chains, 80
signal processing programs run by interneurons and effector neurons, 84–86
signal transfer programs, 80–81
signal transmission between neurons, 82
Network structure, formation of, 87–93
Neural chain links, 83
Neuron activity, restoring, 82–83
Neurons form NS signal chains, 80
Nuclear-cytoplasmic protein transport, 16

Nucleoporins (Nup), 16
Nutrition program, 56–59

O

OMs, *see* Organ makers
Operational neuron (ON), 87
Operational register (OR), 146, 148
OR, *see* Operational register
Organ makers (OMs), 55, 65, 67, 75, 99, 106, 110, 162, 163, 165, 166

P

Pathogen-associated molecular patterns (PAMP), 66
Pattern, 47
PC, *see* Phosphatidylcholine
PCRF, *see* Program of controlling the rotation of flagella
PeCAM makers, 68, 78
Peptidyl-tRNA-binding site, 10
Peroxisomes, 15
Personal information programs, hierarchy of, 147–148
p53 proteins, 38–39, 46, 53, 150, 157
Phenotype, 154
Phosphatidylcholine (PC), 14, 15
Phosphatidylethanolamine (PE), 14, 15
Phosphatidyl inositol (PI), 14, 15
Phosphatidylserine (PS), 14
PI, *see* Phosphatidyl inositol
PL, *see* Program library
Plasma membrane, 15
Program approach
  evolution and, 177–178
  provisions of, 179–184
Program-information approach, time concept in, 136–138
Program-information cell structure, 40–47
Program-information unity of living
  cellular evolution, 154–155
  common program and information structure, 153–154
  evolution and, 177–178
  man, 167–177
  multicellular organisms, 161–167
  unicellular organisms, 155–161
Program library (PL), 123, 124
Program of controlling the rotation of flagella (PCRF), 34
Program scheme of multicellular organisms
  internal and external programs, 55
  main and auxiliary task programs, 55
  MO tasks, 54
  organs and systems, structure of, 55–56
  social multicellular organism structure, 53–54
Program space (PS), 97–98
Program structure, 156–157
Program system (PS), 51, 53, 149, 159, 172, 173
  stochastic degradation in, 139–140
Programtype (protype), 154
Prosecutor's Office, 119, 120
Protective programs, 33, 36–37, 161
Proteins, 3
Provision programs, 11–14
PS, *see* Phosphatidylserine; Program space; Program system
Postsynaptic potential (PSP), 83, 85–86, 93
PVA, *see* Pyruvic acid (PVA)
Pyruvic acid (PVA), 13

R

RanGAP1, 17
RanGDP, 17
RanGTPase, 17
Recombination, 45, 158
Regulatory programs, 130
Reparation, 45
Replication, 45, 158
Reproduction programs, 116
Resource, procurement programs, 36
Resource provision programs, 160
Ribonucleic acid (RNA), 3–7
  matrix, 48
  messenger, 6, 10
  polymerase, 4, 6–8
  ribosomal, 6
  transport transfer, 6, 8
Ribosomal RNA (rRNA), 6
RNA, *see* Ribonucleic acid
rRNA, *see* Ribosomal RNA

S

Sanitary programs, 115
Science, role in program-information unity of living, 173–174
Scientific method, 174
Scientific stage in evolution of volitional programs, 132–140
  applied and fundamental science, 133
  methodology, 133–135
  space, 135–136
  stochastic degradation in program system, combating, 139–140
  stochasticity, 138–139
  time, 135–136
  time concept, in program-information approach, 136–138
SCPs, *see* Systems of a cell program
Search for a food object (SFO), 96
Sensor, 98
Sequential inhibition, 22

Signal marks, 45, 158–159
Signal processing programs, run by interneurons and effector neurons, 84–86
Signal program (SP), 24–27, 36–38, 160
Signal recognition particle (SRP), 17
Signal transfer programs, 80–82
Signal transmission between neurons, 82
Simulation, 148–149
Site reproduction, 14–15
Smooth muscle cells (SMCs), 57–60, 72–74, 86
SMOPs, *see* System of MO programs
SMs, *see* System makers
Socialization, 51
Social multicellular organism structure, 53–54
Social nature of human communities, 168
Social structure of human communities
  autonomous system of volitional programs, 113
  collective programs, types of, 114–116
  human social paradigm, 114
Sorting signals (SSs), 16, 17, 44
Space, 49, 135–136
Spatio-temporal transformations, 48
SP, *see* Signal program; Stimulating program; System of programs
SRP, *see* Signal recognition particle
Stimulating program (SP), 147–148
Stochasticity, 138–139
Stochastic process, and predictable program, 2
Storage devices, 47–48
Supply, procurement programs, 129
Symbol, 145–147
System makers (SMs), 61, 162
System of MO programs (SMOPs), 53, 55
System of programs (SP), 35
Systems of a cell program (SCPs), 53

T

Task functional visual maker (TFVM), 98, 99
Task movement maker (TMM), 101–102
TFE, *see* Time of fixed existence
Time, 49, 135–136
  in program-information approach, 136–138
Time of fixed existence (TFE), 93, 109, 166
Tissue makers (TMs), 162
TMM, *see* Task movement maker
TMs, *see* Tissue makers
Transcription, 45
Transition from unicellular to multicellular organisms, 51–52
Translation, 45
Transport programs, 15–19, 36, 71–78, 129, 160
  blood supply system, 71
  circulation, 75
  heart, 72
  vessel structure/types, 72–74
Transport transfer RNA (tRNA), 6, 8, 156
  eukaryotic, 9
  3′-terminus of, 9
tRNA, *see* Transport transfer RNA

U

Unicellular organisms
  cellular evolution, 161
  complete system of unicellular programs, 159–161
  evolution of, 39–40
  external programs of, 33–35
  full program system of, 35–39
  information network, 157
  information programs, task and structure of, 158–159
  makers, 155–156
  management of programs, 157
  program structure, 156–157
  in volitional programs, inheritance and development of, 149–152

V

Vessel structure/types, 72–74
Volcano, chemical laboratories in, 1–2
Volitional programs (VPs), 112–114, 167, 168
  administrative management and regulatory programs, stages of, 130
  basic consumer programs, 128–129
  collection of information, 126
  collective, 168–169
  communication, 126–127
  energy programs, 129
  formation of information, 126
  history as evolution of, 172–173
  history of mankind as evolution of, 127–132
  image and symbol, 145–147
  information in, 122–127
  information programs, 129–130
  information system, features of, 176–177
  language and, 144–145
  markers, structure of, 127
  market regulatory programs, development of, 130–131
  mechanism of, 175–176
  memory, 123–126
  military programs, 131
  moral regulatory programs, 131
  personal, 168–169
  personal information programs, hierarchy of, 147–148
  processing of information, 126
  scientific stage in evolution of, 132–140
  simulation and manipulative programs, 148–149

Volitional programs (VPs) (*Continued*)
supply, procurement programs, 129
transport programs, 129
unicellular and multicellular information systems, inheritance and development of, 149–152
VPs, *see* Volitional programs

W

World picture, 141–144, 174–175

Z

Zymogen (pro-ferment), 20–21

Printed in the United States
by Baker & Taylor Publisher Services